Advanced Smart Grid Technologies

Advanced Smart Grid Technologies: Industry Applications and Integration Strategies offers a comprehensive exploration of the cutting-edge innovations and integration techniques shaping the future of power systems. This edited volume brings together leading researchers and industry experts to present real-world case studies, emerging technologies, and strategic frameworks that enable efficient, secure, and sustainable energy management.

This book:

- Explores the full spectrum of smart grid technologies, including artificial intelligence (AI), Internet of Things (IoT), machine learning, blockchain, and digital twin concepts.
- Includes real-world applications from sectors such as manufacturing, transportation, utility services, and smart cities.
- Provides insight into practical frameworks for integrating distributed energy resources (DERs), microgrids, and renewable energy systems into existing infrastructures.
- Discusses securing smart grid networks and enhancing resilience against cyber threats and grid failures.
- Analyses international standards, regulatory policies, and smart grid deployment challenges in various regions.
- Provides forward-looking chapters that outline the trajectory of smart grid development and emerging opportunities.

Covering everything from IoT, AI, and blockchain applications to grid automation, cybersecurity, and renewable energy integration, this book serves as a critical reference for professionals, academics, and policymakers involved in the evolution of smart grid systems across global industries.

Advanced Smart Grid Technologies

Industry Applications and Integration Strategies

Edited by
Krishan Arora, Himanshu Sharma, and
Sanjay Kumar

CRC Press
Taylor & Francis Group
Boca Raton London New York

CRC Press is an imprint of the
Taylor & Francis Group, an **informa** business

Designed cover image: Shutterstock

First edition published 2027
by CRC Press
2385 NW Executive Center Drive, Suite 320, Boca Raton FL 33431

and by CRC Press
4 Park Square, Milton Park, Abingdon, Oxon, OX14 4RN

CRC Press is an imprint of Taylor & Francis Group, LLC

ISBN: 9781041173786 (hbk)
ISBN: 9781041173793 (pbk)
ISBN: 9781003689430 (ebk)

DOI: 10.1201/9781003689430

Typeset in Sabon
by Newgen Publishing UK

Contents

Preface

The global energy sector is undergoing a profound transformation driven by the rapid penetration of renewable energy sources, digitalization, electrification of transportation, and the growing demand for reliable, efficient, and sustainable power systems. In this evolving landscape, the concept of the *smart grid* has emerged as a cornerstone for modern power networks, enabling two-way communication, intelligent control, and seamless integration of diverse energy resources. The edited book *Advanced Smart Grid Technologies: Industry Applications and Integration Strategies* is conceived to capture these advancements and present a comprehensive view of next-generation smart grid solutions.

This book aims to bridge the gap between theoretical developments and real-world industrial applications of smart grid technologies. It brings together recent research outcomes, practical case studies, and integration strategies that address key challenges faced by utilities, industries, researchers, and policymakers. The chapters cover a broad spectrum of topics, including advanced sensing and communication infrastructures, artificial intelligence and machine learning for grid operation, cybersecurity and privacy issues, renewable energy and energy storage integration, electric vehicles, microgrids, demand response, power electronics, and standards-based interoperability.

A distinguishing feature of this volume is its strong emphasis on industry-oriented applications and deployment strategies. While many existing works focus primarily on conceptual frameworks, this book highlights implementation aspects, system-level integration, and scalability issues encountered in practical smart grid deployments. Contributions from academia, research institutions, and industry experts provide multidimensional perspectives, making the content both technically rigorous and application-driven.

The book is intended to serve as a valuable reference for postgraduate students, doctoral researchers, practicing engineers, utility professionals, and technology developers working in the domains of smart grids, power systems, and sustainable energy. It may also support curriculum development and act as a guide for future research directions in advanced grid technologies.

We hope that this book will stimulate further innovation, encourage interdisciplinary collaboration, and contribute meaningfully to the advancement and successful integration of smart grid technologies in modern power systems.

About the editors

Krishan Arora is associated with Lovely Professional University as a Professor with more than 17 years of experience in academics and research. He is currently the Head of Department, Power Systems in School of Electronics and Electrical Engineering, Lovely Professional University. He obtained his Ph.D. in Electrical Engineering from MM (Deemed to be) University, Ambala. His research interests include electrical machines, artificial intelligence, load frequency control, electric vehicles, waste management, renewable energy, Internet of Things, and modernization of smart grids. He has published more than 130 research papers in reputed Science Journals and Conferences. He has organized several workshops, summer internships, International Conferences, special sessions, and expert lectures. He has also published 15 Indian patents and 1 copyright. He has supervised 1 PhD scholar, 10 postgraduate scholars, and more than 30 undergraduate students' projects. He has attended/participated in 24 National/International Online webinars. He has edited 25 books in different areas of Electronics and Electrical Engineering. He is associated for editing work with top publishers.

Himanshu Sharma has completed his Ph.D. in the area of power electronics from Punjab Engineering College, Chandigarh. He did his M.Tech in Electrical Engineering from Punjab Engineering College, Chandigarh. He is currently deputed as Dean R&D at G.H. Raisoni College of Engineering and Management, Nagpur, since December 2024. He has published more than 30 research papers in refereed reputed Journals and Conferences. He has organized several workshops, summer internships, International Conferences, and expert lectures for students. He has supervised more than 5 Ph.D. students and more than 15 UG/PG students. He has attended/participated in 20 National/International Online webinars. His area of expertise includes power electronics, waste management, machine learning, and optimization techniques. He has taught various courses at the UG and PG levels, like power electronics, non-conventional energy sources, electric drives, wireless communication, and Internet of Things, etc. He has edited many books for reputed publishers.

Sanjay Kumar is presently working as an Assistant Professor in the University Institute of Technology, Himachal Pradesh University, Shimla, India. He has 7 years of teaching and research experience at NITs and other government institutes. He has published 30 research papers in various International/National Journals and Proceedings of the reputed International/National Conferences. His area of research includes power system, renewable energy, network planning, micro grid, optimization, GIS, and machine learning. He has been granted one international patent, and one national patent has been successfully filed. He has published a number of research papers in various journals and conferences. He has delivered expert lectures in various colleges/universities in India. He is a reviewer member of various international journals and conferences.

Contributors

Krishan Arora
School of Electronics and Electrical Engineering
Lovely Professional University
Jalandhar, Punjab

Abhishek Bajaj
Department of CSE-Applied-Sciences
CGC University, Mohali

Ranjit Kumar Bindal
Department of Electrical Engineering
Chandigarh University, Punjab, India

Rekha Chaudhary
School of Electronics and Electrical Engineering
Lovely Professional University
Jalandhar, Punjab

Sachin Chawla
ECE Department
Chandigarh University, Mohali-140413, Punjab, India

Javed Dhillon
School of Electronics and Electrical Engineering
Lovely Professional University
Jalandhar, Punjab

Karun Handa
School of Computer Science and Engineering
Lovely Professional University
Jalandhar, Punjab

Siga Karthik
School of Electronics and Electrical Engineering
Lovely Professional University
Jalandhar, Punjab

Kirti
ECE Department
Sat Kabir College of Engineering, Bahadurgarh, Haryana

Keshav Kumar
Department of Electronics and Communication Engineering
Pranveer Singh Institute of Technology Kanpur, India

Qutaiba Humadi Mohammed
Media Technology and Communications Engineering Department
College of Engineering

University of Information Technology and Communications, Baghdad, Iraq

Anupama Namburu
School of Engineering
Jawaharlal Nehru University, New Delhi

Irfan Ahmad Pindoo
School of Electronics and Electrical Engineering
Lovely Professional University
Jalandhar, Punjab

Prashant
School of Chemistry
Lovely Professional University
Jalandhar, Punjab

Shippu Sachdeva
School of Electronics and Electrical Engineering
Lovely Professional University
Jalandhar, Punjab

Ramandeep Sandhu
School of Computer Science and Engineering
Lovely Professional University
Jalandhar, Punjab

Prachi Saxena
School of Chemistry
Lovely Professional University
Jalandhar, Punjab

Shahid Shabir
School of Electronics and Electrical Engineering
Lovely Professional University
Jalandhar, Punjab

Akanksha Sharma
School of Electronics and Electrical Engineering
Lovely Professional University
Jalandhar, Punjab

Manish Kumar Sharma
School of Computer Science and Engineering
Lovely Professional University
Jalandhar, Punjab

Gurjot Singh
Department of Electrical Engineering
Chandigarh University, Punjab, India

Pradeep Singh
ABVGIET-Pragatinagar
Shimla, Himachal Pradesh

Ranveer Siyan
Department of Electrical Engineering
Chandigarh University, Punjab, India

Sonia Vatta
Computer Science & Engineering
Rayat Bahra University, Mohali, Punjab, India

Chapter 1

Introduction to smart grids

Siga Karthik and Rekha Chaudhary

1.1 INTRODUCTION

The electric power grid has been one of the most transformative engineering achievements of the last century, enabling the modern industrial, commercial, and digital economies [1]. From the early centralized power plants that lit up small towns to the vast interconnected high-voltage networks supplying megacities, the grid has shaped society and technological progress like few other infrastructures [1].

Over the past century, the grid has grown in size and complexity, but its underlying design philosophy has not really changed one-way electricity flow from large generation stations to passive loads [2]. This architecture worked beautifully during the period of forecastable demand, stable sources of generation, and vertically integrated utilities. Nonetheless, the 21st century has introduced unprecedented challenges: increasing global energy requirements, the imperative of mitigating climate change, the swift growth of distributed and renewable energy sources, transport electrification, and the growing necessity for cyber-physical resilience (Figure 1.1) [3].

Legacy grids, though strong, can't handle these challenges. They have operational limitations in accepting variable renewable sources like solar and wind, lack of real-time view of grid state, sluggish fault detection and restoration mechanisms, large transmission and distribution losses, and few opportunities for active consumer engagement in the energy markets [4]. The outcome is an urgent necessity for a smarter, more flexible, and cleaner power system [5].

This is where the smart grid comes in as a revolutionary solution. A smart grid is not one technology but a sophisticated, networked ecosystem that brings digital sensing, real-time communication, automated control, and distributed intelligence together into the power network [6]. It is a cyber-physical system able to monitor and optimize power flows in real time, balance supply and demand in real time dynamically, self-heal following faults, and actively involve consumers—now frequently prosumers in energy generation, storage, and management (Figure 1.2) [6].

DOI: 10.1201/9781003689430-1

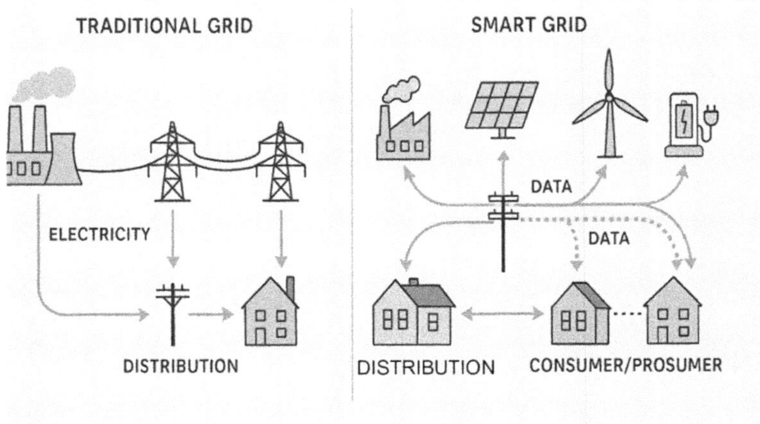

Figure 1.1 Conventional vs. Smart Grid energy flow.

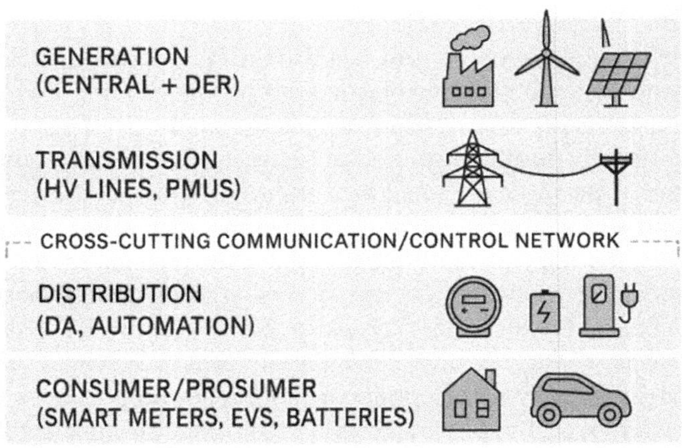

Figure 1.2 Smart grid layered architecture.

Smart grid applications cover the whole electricity value chain. In generation, they enable optimal dispatch of both centralized and distributed energy resources (DERs). In transmission and distribution, they provide wide-area situational awareness with phasor measurement units (PMUs) and intelligent fault location, isolation, and service restoration (FLISR) [7]. For the consumer interface, advanced metering infrastructure (AMI) and home energy management systems (EMS) give users detailed energy data, dynamic pricing, and automation features. Interoperability with new technologies like electric vehicles (EVs), energy storage systems (ESS), and microgrids adds flexibility and reliability [7].

Aside from the physical infrastructure, smart grids utilize contemporary IT technologies such as artificial intelligence (AI), the Internet of Things (IoT), blockchain, and edge computing to provide predictive maintenance, demand response (DR) optimization, renewable forecasting, and secure peer-to-peer (P2P) energy trading [8]. These functionalities not only increase operational efficiency but also enable the global drive toward decarbonization and energy democratization [8].

The smart grid deployment has shown quantifiable results: minimized outage times, enhanced asset utilization, deeper penetration of renewable energy, and more customer participation in energy efficiency [5]. Furthermore, the smart grid revolution is a two-way street: as digital smarts improve the power grid, high-quality electricity provided by advanced grids, which is reliable, is needed to energize the very data centers, communications networks, and control systems that enable the smart grid [1].

This chapter is intended to provide an in-depth introduction to smart grids, following their development, structure, major components, foundational technologies, applications, issues, and emerging trends. By integrating technical observations with practical examples, it functions as a guide for engineers, researchers, policymakers, and industry experts looking to comprehend and partake in the continued transformation of the electric power infrastructure [6].

1.2 EVOLUTION OF THE POWER GRID OVER TIME

The contemporary power grid is the product of more than a century of technological advancements, infrastructure build-out, and operational improvement [9]. Its evolution can be roughly split into four different periods, each with unique technical capabilities, market structures, and operating philosophies [9].

1.2.1 Period I: centralized generation and local distribution (later 19th–early 20th century)

The first electric networks were tiny, isolated grids supplied by central generators and normally serving a single neighborhood or factory. The systems used direct current (DC) generation until the advent of alternating current (AC) transmission, introduced by Nikola Tesla and George Westinghouse, made it possible to deliver efficiently over long distances [10]. Characteristics: Local control, limited range, low redundancy.

Restrictions: Limited economic scalability, no interconnection among systems.

1.2.2 Era II: national and regional grid interconnection (mid-20th century)

It was spurred on by industrial expansion and urbanization that utilities started connecting generation and distribution systems into regional and national grids [11]. Fossil-fuel and hydroelectric generating plants of large sizes provided the supply backbone [11].

Characteristics: Centralized generation, high-voltage AC transmission, vertically integrated utilities.

Benefits: Economies of scale, better reliability through redundancy. Restrictions: Rigid, top-down operation, minimal real-time monitoring, and inflexible to variable renewable sources.

1.2.3 Era III: automation and SCADA systems (late 20th century)

By the 1970s and 1980s, the advent of Supervisory Control and Data Acquisition (SCADA) systems enabled remote monitoring and limited automation of grid infrastructure. Utilities were able to monitor power flows, voltage levels, and breaker statuses in near real time [12].

Key features: Remote control of substations, basic load balancing, better outage management.

Limitations: Yet still largely centralized, low integration of renewable energy, little consumer interaction.

1.2.4 Era IV: digitalization and emergence of the smart grid (21st century)

Sensors, communications infrastructure, and computing advancements have made possible smart grids decentralized, digitally networked systems that can operate dynamically [13]. This paradigm brings DERs, renewable generation, AMI, wide-area monitoring systems (WAMS), and cybersecurity systems together [13].

Major characteristics: Two-way power and data exchanges, predictive analytics, self-healing.

Benefits: Greater renewable penetration, consumer engagement, enhanced resilience (Figure 1.3).

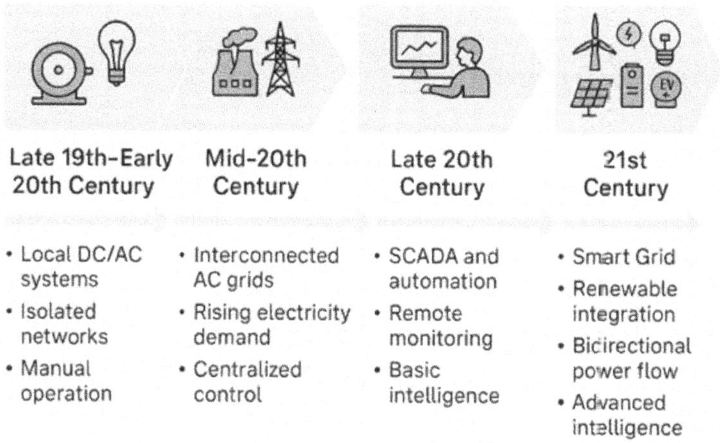

Figure 1.3 Timeline of power grid evolution.

1.3 SMART GRIDS MOTIVATION

The evolution of conventional electricity networks into smart grids is not a matter of choice—it is a necessity based on the increasing complexity, exposure, and sustainability demands of contemporary power systems [14]. Several converging drivers are transforming the global energy framework, rendering traditional grid structures increasingly unsuitable [14].

1.3.1 Renewable energy source integration

Utilities and governments all over the world are signing up to aggressive decarbonization commitments to combat climate change [15]. To meet these objectives, there needs to be mass deployment of renewable energy technologies like solar photovoltaics (PV), wind turbines, and small hydro.

The intermittent and variable nature of renewables creates uncertainty and stress on grid stability and reliability. Smart grids offer the sensing, prediction, and control functions needed to manage generation and demand dynamically in the face of varying renewable output [15].

1.3.2 Growing demand and electrification

World electricity demand is expected to increase strongly as a result of economic development, urbanization, and the electrification of transport, heat, and industry [16]. Conventional grids, developed for static load patterns, are not well prepared to meet these fast-evolving consumption profiles without extensive modernization [16].

1.3.3 Aging infrastructure

Much of the global transmission and distribution infrastructure is decades old. Aging transformers, lines, and substations increase losses, failure rates, and maintenance costs [17]. Smart grids enable condition-based monitoring and predictive maintenance to extend asset life and optimize replacement schedules [17].

1.3.4 Resilience against disruptions

Severe weather, natural catastrophes, and cyber-attacks are on the rise in frequency and intensity [17]. Conventional grids do not have the self-healing and islanding features to restore quickly. Smart grids isolate faults, re-route power flows, and continue supply to essential loads in case of emergency [17].

1.3.5 Consumer empowerment

Customers are not only passive recipients of electricity anymore; they are now prosumers, producing and storing electricity through rooftop solar, home batteries, and EVs [18]. Smart grids allow two-way energy flows, real-time prices, and active engagement in DR programs, which enable customers to actively participate in the operation of the grid (Figure 1.4) [18].

1.4 DEFINITION AND FUNDAMENTAL PRINCIPLES OF SMART GRIDS

Although the smart grid term has become very much in vogue in academic, business, and policy discourse, its meaning will depend on the point of view—utility, regulator, technology vendor, or consumer [19]. Essentially, a smart grid is an upgraded electricity system that embeds advanced sensing, communications, automation, and analytics technologies into generation, transmission, distribution, and use of electricity to enhance efficiency, reliability, sustainability, and security [19].

1.4.1 Key organizations' definitions

1.4.1.1 U.S. Department of Energy (DOE)

An electricity delivery system that uses digital technology to monitor and control the transportation of electricity from all sources of generation to satisfy the diverse electricity needs of end-users [20].

Figure 1.4 Drivers for smart grid development.

1.4.1.2 International Electrotechnical Commission (IEC)

A smart grid is a power network that can manage the behaviour of all connected user's generators, consumers, and prosumers to provide sustainable, economic, and secure electricity supplies in an efficient way [20].

1.4.1.3 CIGRE (International Council on Large Electric Systems)

A smart grid is a harmonious fusion of conventional power system infrastructure with modern information and communication technology (ICT) for real-time monitoring, predictive control, and optimised operation.

These definitions have common components: intelligence, two-way communication, integration of distributed resources, and consumer participation [21].

1.4.2 Smart grids core principles

Observability: The grid shall offer operators high-resolution, real-time visibility into voltages, currents, frequencies, power quality, and equipment status throughout the network [22].

Controllability: The capacity to impact system behavior through automated control actions, DR signals, and dispatch of distributed generation [22].

Interoperability: Efficient communication and data exchange among devices, systems, and organizations by means of standardized protocols (e.g., IEC 61850, DNP3, DLMS/COSEM).

Scalability: Ability to support increasing demand, additional DERs, and emerging technologies without a need for basic redesign [22].

Resilience: The ability to resist and recover rapidly from physical faults, cyber-attacks, and natural disasters.

Sustainability: Incorporating renewable energy and storage to decrease carbon emissions and dependence on fossil fuels.

Security: safeguarding both information technology (IT) and operational technology (OT) layers against cyber-physical attacks (Figure 1.5) [22].

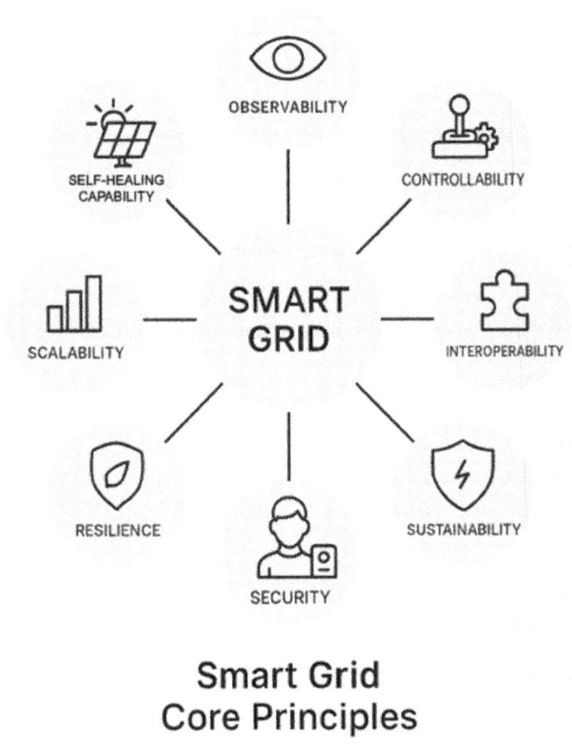

Figure 1.5 Smart grid core principles wheel.

1.5 KEY FEATURES OF SMART GRIDS

Smart grids are differentiated from conventional electricity grids not just by the integration of digital elements, but by a group of operational and functional features that support an intelligent, adaptive, and participative power system. These features determine the smart grid's capacity to address the changing needs of current energy systems [23].

1.5.1 Two-way power and communication flow

Unlike conventional grids, where there is one-way electricity flow from the centralized generation to the consumers, Smart grids facilitate two-way flows of data and electricity [24]. This makes it possible for DERs like roof-top solar, small wind turbines, and battery storage systems to supply power into the grid while sharing operational data with the utility control centers [24].

1.5.2 Real-time monitoring and control

Smart grids utilize high-tech sensing equipment like PMUs, intelligent electronic devices (IEDs), and smart meters to monitor system conditions in real time. Data is pushed into EMS and distribution management systems (DMS) for quick decision-making and automated management.

1.5.3 Distributed generation integration

Smart grids are meant to integrate decentralized and renewable energy sources in a seamless manner. This involves managing their variability using forecasting, load balancing, and storage coordination [25].

1.5.4 Self-healing capability

By using automation, intelligent FLISR, smart grids can automatically detect, localize, and isolate faults, redirecting power flows to shorten outage times without the need for human intervention [25].

1.5.5 Demand-side participation

Customers are themselves involved in energy management through DR schemes, dynamic tariffs, and home EMS. This makes passive customers active market players [26].

1.5.6 Cybersecurity enhancement

With the expansion of communication networks into the power system, smart grids have multi-layered cybersecurity measures to safeguard operational assets as well as information assets [27] (Table 1.1; Figure 1.6).

Create a hexagon or circular infographic with "Smart Grid" in the center. Around it, place labeled icons for: Two-Way Power/Data Flow (arrows in both directions), Real-Time Monitoring (gauge/meter icon), Distributed Generation Integration (solar/wind icons), Self-Healing (broken line auto-reconnect), Demand-Side Participation (user + smart meter), Enhanced Cybersecurity (shield/lock) [27]. Use connecting lines from each icon to the central "Smart Grid" label. Include a small caption under each icon summarizing the benefit.

1.6 SMART GRID ARCHITECTURE

A smart grid architecture is the unification of physical power infrastructure and digital technologies for communication, control, and automation. It covers the entire value chain of electricity from generation to end-use consumption with a combination of centralized and distributed assets [28].

A smart grid may be described with a layer-based approach, with each layer having particular functions but being connected with the others with an integrated communication and control system [29].

1.6.1 Generation layer

This layer has both centralized power generation plants (coal, gas, hydro, nuclear) and DERs like solar PV, wind power plants, biomass facilities, and small hydro. Smart grid technologies enable maximum dispatch, renewable forecasting, and storage coordination to maintain system stability [30].

Table 1.1 Comparison of traditional grid and smart grid characteristics

Aspect	*Traditional grid*	*Smart grid*
Power flow	One-way	Two-way (centralized + distributed)
Monitoring	Manual/periodic	Real-time, continuous
Fault management	Manual detection and switching	Automated self-healing (FLISR)
Renewable integration	Limited	High penetration with storage and forecasting
Consumer eole	Passive	Active (prosumers, DR participation)
Communication	Minimal	Integrated, standardized, interoperable
Cybersecurity	Perimeter-based	Multi-layered, proactive

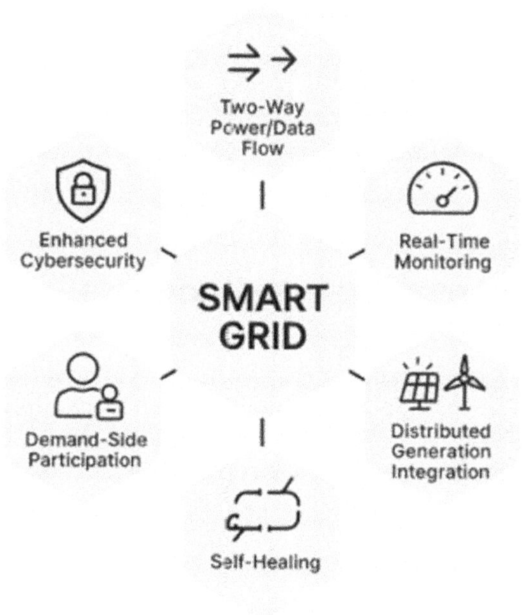

Figure 1.6 Smart grid key characteristics infographic.

1.6.2 Transmission layer

High-voltage transmission lines and substations transmit electricity over long distances. In smart grids, the transmission layer is augmented with PMUs and WAMS, offering synchronized, real-time visibility of grid conditions for stability control [30].

1.6.3 Distribution layer

At medium- and low-voltage levels, DA supports smart FLISR, voltage control, and load balancing. Sensors, smart switches, and automated reclosers enhance efficiency and reliability [31].

1.6.4 Consumer/prosumer layer

The consumer tier now encompasses prosumers who produce, store, and even sell electricity back to the grid. AMI and home EMS provide consumers with detailed information on their energy consumption and allow them to participate in DR programs [31].

1.6.5 Communication and control layer (cross-cutting)

Concurrently running across all physical layers is the communication and control infrastructure [32]. This comprises wired (fiber optic, power line communication [PLC]) and wireless (radio frequency [RF] mesh, LTE/5G) communications and standardized protocols such as IEC 61850, DNP3, and DLMS/COSEM. DMS and centralized EMS communicate with edge computing nodes, local controllers, and aggregators (Figure 1.7).

1.7 KEY ELEMENTS AND TECHNOLOGIES OF A SMART GRID

The smart grid is not a piece of hardware or a system but an ecosystem of interrelated elements and technologies that together provide real-time monitoring, adaptive control, renewable integration, and consumer engagement [33]. All the elements have a vital role to enhance the performance, resilience, and sustainability of the grid [33].

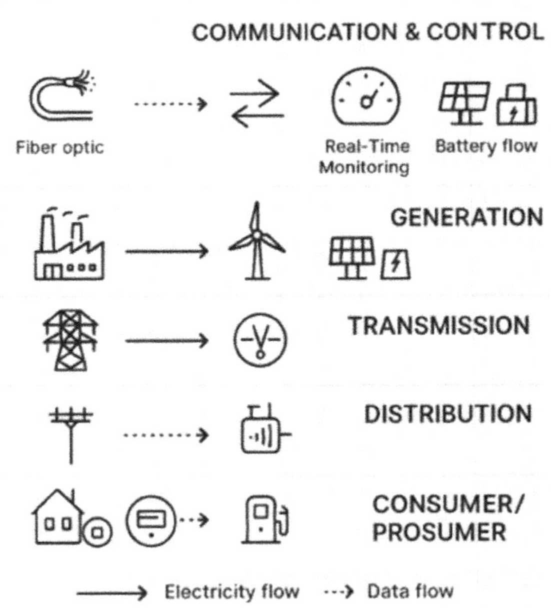

Figure 1.7 Layered smart grid architecture.

1.7.1 AMI

AMI includes smart meters, communication networks, and data management systems offering two-way communication between consumers and utilities.

Functions: Remote meter reading, detection of outages, dynamic pricing, load profiling.

Benefit: Enhances billing accuracy, DR programs, and consumer participation [34].

1.7.2 PMUs

PMUs are real-time high-speed sensors measuring voltage, current, and phase angle through GPS-synchronized data.

Functions: Wide-area situational awareness, monitoring of stability, detection of oscillations.

Benefit: Facilitates early grid disturbance detection and enables real-time control.

1.7.3 Distribution automation (DA) systems

DA comprises smart switches, reclosers, and sensors placed in distribution lines to automate fault detection, isolation, and restoration of service (FLISR) [35].

Functions: Voltage control, load balancing, self-healing operation.

Benefit: Shortens outage duration and enhances service reliability.

1.7.4 EMS and DMS

EMS controls generation and transmission operations, whereas DMS optimizes medium- and low-voltage distribution systems.

Functions: Forecasting load, network reconfiguration, voltage/volt ampere reactive (VAR) control [36].

Benefit: Enhances operational effectiveness and asset utilization.

1.7.5 DERs

DERs comprise rooftop solar PV, wind turbines of small capacity, biomass plants, microturbines, and fuel cells [36].

Functions: Generation that is decentralized near the load centers to lower transmission losses.

Benefit: Increases sustainability and energy independence.

1.7.6 ESS

Battery energy storage, pumped hydro, and flywheels mitigate supply-demand mismatches.

Functions: Frequency regulation, peak shaving, renewable firming.

Benefit: Enhances reliability and renewable integration [37].

1.7.7 EV integration

EVs are both loads and potential energy storage resources (vehicle-to-grid, V2G).

Functions: Grid stabilization, use of renewable energy [37].

Benefit: Facilitates decarbonization of transport and grid flexibility.

1.7.8 Cybersecurity systems

Smart grids need to be protected by multi-layered cybersecurity measures on OT and IT.

Functions: Detection of intrusions, encryption, access control, anomaly detection.

Benefit: Protects against cyber-physical threats (Figure 1.8) [38].

1.8 BENEFITS OF SMART GRIDS

Implementation of smart grid technologies provides benefits in technical, economic, environmental, and societal terms [39]. The benefits are not only felt by utilities but also by consumers, regulators, and the wider economy.

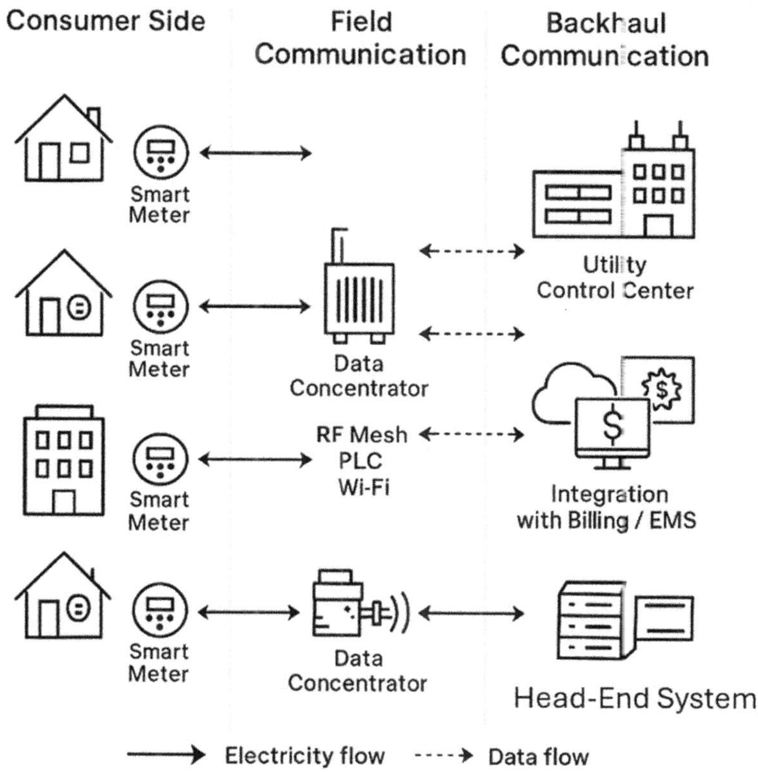

Figure 1.8 Advanced metering infrastructure (AMI) network topology.

1.8.1 Operation benefits

Enhanced reliability: Automation, self-healing, and real-time monitoring minimize outage frequency and duration [40].

Better power quality: Voltage regulation, harmonic control, and frequency stability enhance service quality [40].

Effective use of assets: Condition-based maintenance and load balancing lengthen equipment life and save capital costs [40].

1.8.2 Economic benefits

Lower cost of operations: Remote fault management, predictive maintenance, and automated metering decrease labor and maintenance expenses [41].

Dynamic pricing and DR: Promotes consumers to change usage to off-peak times, reducing peak demand charges [41].

Deferral of infrastructure investment: Improved use of installed assets delays expensive capacity additions [41].

1.8.3 Environmental benefits

Lower carbon emissions: More integration of renewables lowers dependence on fossil fuels.

Energy efficiency: Live data on energy consumption inspires conservation on both the utility and consumer sides [42].

Less transmission losses: Distributed generation closer to points of consumption reduces wastage of energy [42].

1.8.4 Societal benefits

Consumer empowerment: Prosumers are able to produce, store, and sell energy, making more people part of the energy market.

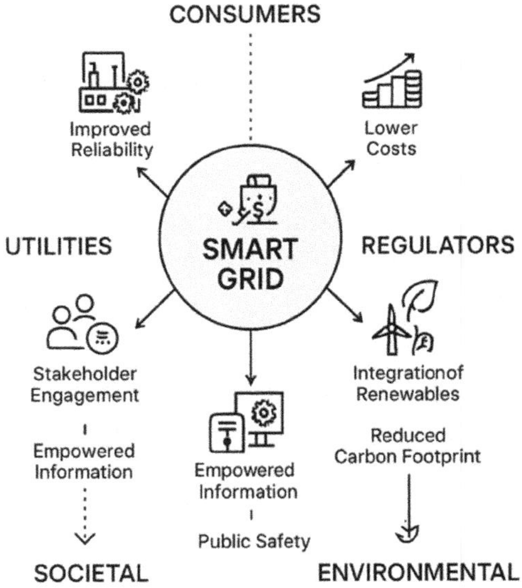

Figure 1.9 Smart grid benefits by stakeholder.

Resistance to disasters: Quick recovery from natural disasters and cyber-attacks ensures community stability [42].

Job creation: The installation of smart grid infrastructure offers opportunities in manufacturing, IT, and engineering (Figure 1.9).

1.9 APPLICATIONS OF SMART GRIDS

Smart grids are not simply a 21st-century upgrade to infrastructure; they support new models of operations, market mechanisms, and services throughout the energy system [43]. These applications take advantage of the smart grid's intelligence, connectivity, and adaptability to meet various technical and societal requirements [43].

1.9.1 Microgrids and islanding

Smart grids facilitate microgrid operation through localized energy infrastructure that may operate in parallel or isolated from the primary grid [44].

Applications: Military installations, campuses, industrial parks, isolated communities.

Benefits: Back-up power resilience during outages, integration with renewable resources, lower transmission losses [44].

1.9.2 DR

DR programs enable utilities to manage consumer energy consumption in real-time by providing incentives for shifting loads [45].

Applications: Managing peak load reduction, managing variable renewable output balancing.

Benefits: Prevents expensive peaking plants, enhances grid stability [45].

1.9.3 Renewable energy integration

Smart grids enable forecasting, scheduling, and balancing of renewable sources like wind and solar [46].

Applications: Real-time prediction of output, minimization of curtailment, hybrid renewable-storage facilities.

Advantages: Enhances renewable penetration without compromising on reliability [46].

1.9.4 EV integration

Charging systems of EVs are coordinated with grid conditions by smart charging systems, and stored energy in EVs can be sold back to the grid by V2G technology [46].

Applications: Shaving of peaks, frequency regulation.

Advantages: Enhances grid flexibility and enables clean transport.

1.9.5 P2P energy trading

Blockchain-based smart grids enable consumers and prosumers to exchange electricity P2P without central intermediaries [47].

Applications: Neighborhood energy exchanges, community solar projects.

Benefits: Fosters local renewable adoption and market engagement.

1.9.6 Fault detection and self-healing advanced

Automation and data analytics facilitate quick detection and separation of faults, reducing outage duration [47].

Applications: High-renewable grids, distribution networks.

Benefits: Enhances reliability and shortens restoration times (Figure 1.10).

1.10 SMART GRID IMPLEMENTATION CHALLENGES

Although smart grids hold a lot of promise, large-scale implementation is threatened by a variety of challenges across technical, economic, regulatory, and social fronts. The removal of these hurdles is necessary for the realization of the full potential of smart grid technologies [48].

1.10.1 Technical challenges

Interoperability challenges: Heterogeneous devices and systems from various vendors need to communicate seamlessly with each other, necessitating compliance with open standards.

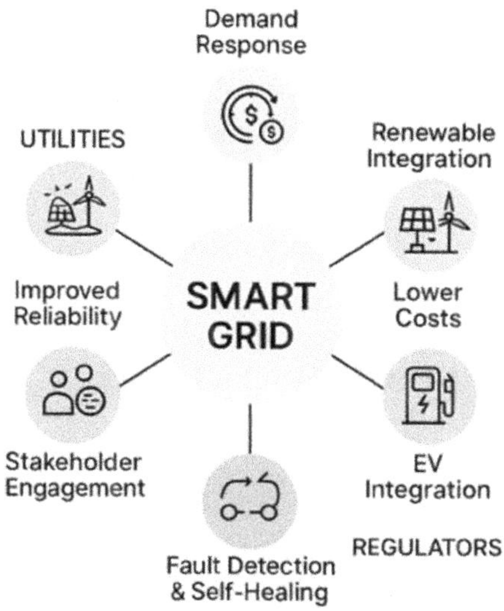

Figure 1.10 Smart grid application domains.

Integration of legacy infrastructure: Legacy power infrastructure was not optimized for bidirectional power flows or real-time monitoring.

Data management complexity: The enormous quantity, speed, and diversity of grid data need strong storage, processing, and analytics functions.

Cybersecurity risks: Greater connectivity provides new avenues of attack for hackers.

1.10.2 Economic challenges

High initial investment: Installation of sensors, communications networks, and control systems requires high capital investment.

Uncertain return on investment (ROI): Some benefits like enhanced reliability and loss reduction are sometimes difficult to determine financially [48].

Cost recovery mechanisms: Utilities need facilitating tariff structures to get back smart grid investments without overloading consumers.

1.10.3 Regulatory and policy challenges

Lack of standardization: Lack of widely accepted technical standards postpones interoperability.

Data privacy concerns: Consumption data from real-time creates consumer privacy and data ownership issues [49].

Slow policy adaptation: Policy structures are slow to adapt to technological advancements.

1.10.4 Social and behavioral challenges

Consumer awareness: Consumers are mostly unaware of smart grid features or likely advantages.

Acceptance of dynamic pricing: Customers oppose variable tariffs even though they have economic and environmental advantages.

Trust in technology: Worries regarding dependability, information security, and equity can delay usage (Figure 1.11) [50].

1.11 FUTURE DIRECTIONS AND TRENDS IN SMART GRIDS

The future smart grid will be autonomous, decentralized, and smart, with the ability to self-optimize in real time under system conditions, market signals, and environmental limitations. Current research and new technologies will continue to build on the capabilities of today's power systems.

1.11.1 Digital twin technology

Virtual copies of grid assets and systems are created to allow for simulation, testing, and optimization prior to implementation [50].

Benefit: Predictive maintenance, operational optimization, and quick fault diagnosis.

1.11.2 AI and machine learning

AI will be more deeply ingrained in control centers, field devices, and DERs for the purposes of load forecasting, fault prediction, and optimal dispatch.

Benefit: Enhances decision-making accuracy and response times [50].

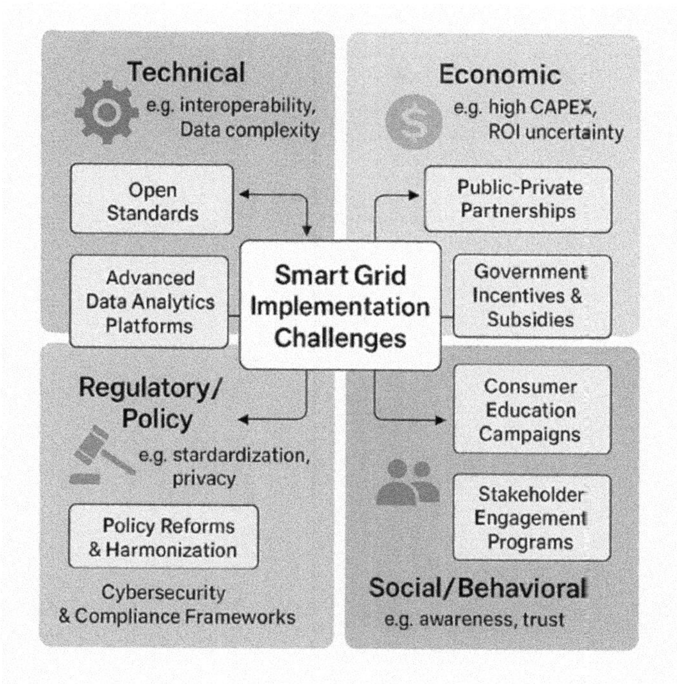

Figure 1.11 Smart grid challenges and mitigation.

1.11.3 Blockchain for energy transactions

Decentralized ledgers will ensure secure P2P energy trading and transparent settlement mechanisms.

Benefit: Removes intermediaries, allowing prosumer-driven markets.

1.11.4 Advanced energy storage

Future batteries (solid-state, flow batteries) and hybrid storage solutions will solve renewable intermittency issues [51].

Benefit: More renewable integration and grid flexibility.

1.11.5 Quantum-safe cybersecurity

As quantum computing begins, post-quantum cryptography will secure smart grids against future cyber-attacks.

Benefit: Provides long-term data security and grid reliability.

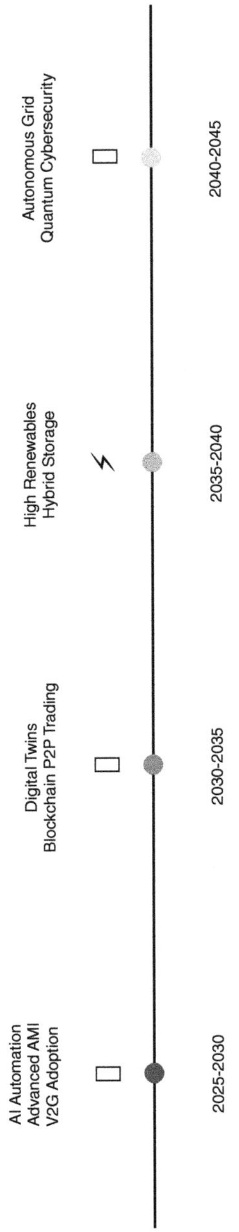

Figure 1.12 Roadmap for the future smart grid.

1.11.6 Vehicle-to-everything (V2X) integration

In addition to V2G, EVs will communicate with buildings, microgrids, and public infrastructure [52].

Benefit: Widens grid support capabilities and energy-sharing possibilities (Figure 1.12).

1.13 CONCLUSION

The evolution of the conventional power grid to a smart grid is a paradigm change in the generation, transmission, distribution, and usage of electricity. By integrating intelligence, communication, and automation into the grid infrastructure, smart grids deal with issues of reliability, sustainability, and consumer empowerment in an increasingly dynamic energy environment.

The deployment of smart grids is not without challenges high capital expenditures, interoperability, cybersecurity, and gaps in policies but the potential advantages will far exceed the obstacles. Efficient deployment needs cooperation among utilities, policymakers, technology suppliers, and consumers, facilitated by open standards, sound cybersecurity frameworks, and dynamic regulation.

With the adoption of renewable energy fast-tracking, EVs going mainstream, and energy markets trending toward decentralization, smart grids will be the enabling catalyst of a cleaner, more efficient, and more secure power system. Added to this fusion will be AI, blockchain, digital twins, and advanced storage technology only to enhance this vision for an autonomous, self-optimizing grid in the coming decades.

REFERENCES

[1] Fang, X., Misra, S., Xue, G., & Yang, D. (2011). Smart grid—The new and improved power grid: A survey. *IEEE Communications Surveys & Tutorials*, *14*(4), 944–980.

[2] Brown, R. E. (2008, July). Impact of smart grid on distribution system design. In *2008 IEEE Power and Energy Society General Meeting-Conversion and Delivery of Electrical Energy in the 21st Century* (pp. 1–4). IEEE.

[3] Arghandeh, R., Von Meier, A., Mehrmanesh, L., & Mili, L. (2016). On the definition of cyber-physical resilience in power systems. *Renewable and Sustainable Energy Reviews*, *58*, 1060–1069.

[4] Amin, S. M., & Wollenberg, B. F. (2005). Toward a smart grid: Power delivery for the 21st century. *IEEE Power and Energy Magazine*, *3*(5), 34–41.

[5] Simões, M. G., Roche, R., Kyriakides, E., Suryanarayanan, S., Blunier, B., McBee, K. D., ..., & Miraoui, A. (2012). A comparison of smart grid

technologies and progresses in Europe and the US. *IEEE Transactions on Industry Applications*, *48*(4), 1154–1162.

[6] Yu, X., & Xue, Y. (2016). Smart grids: A cyber–physical systems perspective. *Proceedings of the IEEE*, *104*(5), 1058–1070.

[7] Parhizi, S., Lotfi, H., Khodaei, A., & Bahramirad, S. (2015). State of the art in research on microgrids: A review. *IEEE access*, *3*, 890–925.

[8] Andoni, M., Robu, V., Flynn, D., Abram, S., Geach, D., Jenkins, D., ... & Peacock, A. (2019). Blockchain technology in the energy sector: A systematic review of challenges and opportunities. *Renewable and Sustainable Energy Reviews*, *100*, 143–174.

[9] Grigsby, L. L. (2018). *The electric power engineering handbook-five volume set*. CRC Press.

[10] Hughes, T. P. (1993). *Networks of power: electrification in Western society, 1880–1930*. JHU Press.

[11] Ilic, M., Galiana, F., & Fink, L. (1998). *Power systems restructuring [electronic resource]: engineering and economics*. https://doi.org/10.1007/978-1-4757-2883-5

[12] Northcote-Green, J., & Wilson, R. G. (2017). *Control and automation of electrical power distribution systems*. CRC press.

[13] Vaccaro, A., Popov, M., Villacci, D., & Terzija, V. (2010). An integrated framework for smart microgrids modeling, monitoring, control, communication, and verification. *Proceedings of the IEEE*, *99*(1), 119–132.

[14] Lasseter, R. H. (2011). Smart distribution: Coupled microgrids. *Proceedings of the IEEE*, *99*(6), 1074–1082.

[15] Connolly, D., Lund, H., Mathiesen, B. V., & Leahy, M. (2010). A review of computer tools for analysing the integration of renewable energy into various energy systems. *Applied Energy*, *87*(4), 1059–1082.

[16] IEA, P. (2022). *World energy outlook 2022*. International Energy Agency (IEA).

[17] Panteli, M., & Mancarella, P. (2015). Influence of extreme weather and climate change on the resilience of power systems: Impacts and possible mitigation strategies. *Electric Power Systems Research*, *127*, 259–270.

[18] Di Bella, A., & Colelli, F. P. (2025). Mitigation strategies can alleviate power system vulnerability to climate change and extreme weather: A case study on the Italian grid. *Environmental Research: Infrastructure and Sustainability*, *5*(1), 015003.

[19] Erol-Kantarci, M., & Mouftah, H. T. (2014). Energy-efficient information and communication infrastructures in the smart grid: A survey on interactions and open issues. *IEEE Communications Surveys & Tutorials*, *17*(1), 179–197.

[20] Ton, D. T., Wang, W. M., & Wang, W. P. (2011, July). Smart Grid R&D by the US Department of Energy to optimize distribution grid operations. In *2011 IEEE Power and Energy Society General Meeting* (pp. 1–5). IEEE.

[21] Gopstein, A., Nguyen, C., O'Fallon, C., Hastings, N., & Wollman, D. (2021). *NIST framework and roadmap for smart grid interoperability standards, release 4.0* (Vol. 10). Department of Commerce. National Institute of Standards and Technology.

[22] Phadke, A. G., & Thorp, J. S. (2008). *Synchronized phasor measurements and their applications* (Vol. 1, No. 2017). Springer.

[23] Pipattanasomporn, M., Feroze, H., & Rahman, S. (2009, March). Multi-agent systems in a distributed smart grid: Design and implementation. In *2009 IEEE/PES Power Systems Conference and Exposition* (pp. 1–8). IEEE.

[24] Lu, X., Wang, W., & Ma, J. (2013). An empirical study of communication infrastructures towards the smart grid: Design, implementation, and evaluation. *IEEE Transactions on Smart Grid*, *4*(1), 170–183

[25] Phadke, A. G. (2002). Synchronized phasor measurements in power systems. *IEEE Computer Applications in Power*, *6*(2), 10–15.

[26] Farhangi, H. (2009). The path of the smart grid. *IEEE Power and Energy Magazine*, *8*(1), 18–28.

[27] Yan, Y., Qian, Y., Sharif, H., & Tipper, D. (2012). A survey on smart grid communication infrastructures: Motivations, requirements and challenges. *IEEE Communications Surveys & Tutorials*, *15*(1), 5–20.

[28] Kuzlu, M., Pipattanasomporn, M., & Rahman, S. (2014). Communication network requirements for major smart grid applications in HAN, NAN and WAN. *Computer Networks*, *67*, 74–88.

[29] Hatziargyriou, N. (Ed.). (2014). *Microgrids: architectures and control*. Wiley.

[30] Phadke, A. G., Thorp, J. S., & Karimi, K. J. (2007). State estimlatjon with phasor measurements. *IEEE Transactions on Power Systems*, *1*(1), 233–238.

[31] Bollen, M. H., & Hassan, F. (2011). *Integration of distributed generation in the power system*. Wiley.

[32] Usman, A., & Shami, S. H. (2013). Evolution of communication technologies for smart grid applications. *Renewable and Sustainable Energy Reviews*, *19*, 191–199.

[33] Sendin, A., Matanza, J., & Ferrús, R. (2021). *Smart Grid Telecommunications: fundamentals and technologies in the 5G Era*. Wiley.

[34] Muqeet, H. A., Liaqat, R., Jamil, M., & Khan, A. A. (2023). A state-of-the-art review of smart energy systems and their management in a smart grid environment. *Energies*, *16*(1), 472.

[35] Phadke, A. G., & Thorp, J. S. (2006, October). History and applications of phasor measurements. In *2006 IEEE PES Power Systems Conference and Exposition* (pp. 331–335). IEEE.

[36] Kezunovic, M. (2011). Smart fault location for smart grids. *IEEE Transactions on Smart Grid*, *2*(1), 11–22.

[37] Katiraei, F., Iravani, R., Hatziargyriou, N., & Dimeas, A. (2008). Microgrids management. *IEEE Power and Energy Magazine*, *6*(3), 54–65.

[38] Vasilakis, A., Zafeiratou, I., Lagos, D. T., & Hatziargyriou, N. D. (2020). The evolution of research in microgrids control. *IEEE Open Access Journal of Power and Energy*, *7*, 331–343.

[39] Ipakchi, A., & Albuyeh, F. (2009). Grid of the future. *IEEE Power and Energy Magazine*, *7*(2), 52–62.

[40] Kundur, P. (2007). Power system stability. *Power System Stability and Control*, *10*(1), 7–1.

[41] Albadi, M. H., & El-Saadany, E. F. (2007, June). Demand response in electricity markets: An overview. In *2007 IEEE Power Engineering Society General Meeting* (pp. 1–5). IEEE.

[42] Lopes, J. P., Hatziargyriou, N., Mutale, J., Djapic, P., & Jenkins, N. (2007). Integrating distributed generation into electric power systems: A review of drivers, challenges and opportunities. *Electric Power Systems Research*, *77*(9), 1189–1203.

[43] Hossain, M. R., Oo, A. M., & Ali, A. S. (2013). Smart grid. In *Smart grids: opportunities, developments, and trends* (pp. 23–44). Springer London.

[44] Lasseter, R. H. (2002). MicroGrids. In *IEEE power engineering society winter meeting* (Vol. 1, pp. 305–308). Scientic Research.

[45] Gellings, C. W. (2005). The concept of demand-side management for electric utilities. *Proceedings of the IEEE*, *73*(10), 1468–1470.

[46] Ackermann, T., Andersson, G., & Söder, L. (2001). Distributed generation: A definition. *Electric Power Systems Research*, *57*(3), 195–204.

[47] Li, Z., Kang, J., Yu, R., Ye, D., Deng, Q., & Zhang, Y. (2017). Consortium blockchain for secure energy trading in industrial internet of things. *IEEE Transactions on Industrial Informatics*, *14*(8), 3690–3700.

[48] Baker, L. (2018). Of embodied emissions and inequality: Rethinking energy consumption. *Energy Research & Social Science*, *36*, 52–60.

[49] Yan, Y., Qian, Y., Sharif, H., & Tipper, D. (2012). A survey on cyber security for smart grid communications. *IEEE Communications Surveys & Tutorials*, *14*(4), 998–1010.

[50] Rahimi, F., & Ipakchi, A. (2010). Demand response as a market resource under the smart grid paradigm. *IEEE Transactions on Smart Grid*, *1*(1), 82–88.

[51] Blaabjerg, F., & Ma, K. (2013). Future on power electronics for wind turbine systems. *IEEE Journal of Emerging and Selected Topics in Power Electronics*, *1*(3), 139–152.

[52] Madni, A. M., Madni, C. C., & Lucero, S. D. (2019). Leveraging digital twin technology in model-based systems engineering. *Systems*, *7*(1), 7.

Chapter 2

Privacy and data security in AI systems

Irfan Ahmad Pindoo

2.1 INTRODUCTION

The convergence of artificial intelligence (AI), robotics, and cognitive systems has revolutionized modern technology. Cognitive machines—robots endowed with AI-driven perception, learning, and decision-making—are increasingly deployed in domains from healthcare and transportation to homes and industry. These AI–robotics integrations promise enhanced autonomy and efficiency, but they also introduce new challenges in privacy and data security. As robots and AI agents work symbiotically, they continuously sense, process, and exchange data (often about humans and their environments), raising concerns about how that data is protected (Eleshin et al., 2024). In fact, the same features, which enable cognitive systems to be powerful ubiquitous sensing, real-time decision-making, and adaptive learning, can be exploited by cyberattacks or invasion of privacy unless they are secured well (Liu et al., 2021; Mestari et al., 2023; Saeed & Alsharidah, 2024; Shahriar et al., 2023).

Privacy in intelligent systems typically means that personal or sensitive information is not accessed or disclosed by an unauthorized party, whereas security means that integrity, availability, and confidentiality of the system data and functionality are preserved (Neupane et al., 2024). The two are both basic to the trust of users. An invasion of privacy (such as a service robot accidentally capturing the chat of a user), or a security failure (such as a hacked autonomous vehicle), can destroy the trust of the population and result in physical injury. With the increasing complexity of AI and its deep integration, it is crucial to make sure that such systems process the information that is consistent with privacy rights and that they can withstand attack. As an example, personal information sharing between humans and robots also requires the strict compliance with the principles of data protection laws such as the California Consumer Privacy Act (CCPA) and the General Data Protection Regulation of the EU (GDPR). To not do so not only can involve legal consequences, but may also cause psychological, physical or loss of autonomy to users communicating with these systems.

DOI: 10.1201/9781003689430-2

Stated differently, strong privacy and security are not additional features but essential facilitators of the secure and ethical implementation of cognitive machines in the society (Liu et al., 2020; Tayyab et al., 2023).

2.2 FOUNDATIONS OF PRIVACY AND SECURITY IN AI SYSTEMS

2.2.1 Definition of privacy and security in the AI context

Within the realm of AI systems, privacy usually means the safety of the personal data of individuals and freedom to control the way the information about the latter is gathered, processed, and distributed. It includes such notions as anonymity (the fact that the person cannot be identified based on the data), confidentiality (access to the data is only granted to the licensed parties), and informed consent (users are aware of and consent to data practice). Indicatively, privacy in an AI-driven healthcare assistant implies that medical and biometric information of a patient remains confidential, and it is used only for the desired purposes, with the consent of the patient and according to the regulations. Security, on the other hand, is what the AI system is doing and what qualities it possesses to protect itself against malicious interference or misuse. This includes ensuring the *integrity* of the system (data and models cannot be tampered with undetected), *availability* (the system and its data remain accessible and functional to authorized users), and continued *confidentiality* of data in the face of cyber threats. In simpler terms, if privacy is about "what data is collected and with whom it is shared," security is about "making sure only the right people (or systems) can access and modify that data." Both are closely intertwined: a security breach can lead to a privacy violation (e.g., a hacker stealing personal data from an AI system), and overly lax privacy practices (e.g., indiscriminate data sharing) can undermine security by broadening the attack surface (Feretzakis et al., 2024). Table 2.1 delineates the comparison of privacy (focus on personal data, anonymity, consent) and security (system integrity, availability, and malicious resilience) aspects in AI systems, highlighting differences in goals, threats, techniques, and regulatory roles.

In AI and robotics, these definitions take on particular nuances. AI systems often *learn* from data, meaning they internalize information from possibly millions of individuals. Ensuring privacy means that an AI model should not reveal sensitive specifics about any one individual from its training data (Liao et al., 2025). Security means that adversaries should not be able to compromise the model or the data pipeline to alter the AI's behavior or extract confidential information. Notably, the *statistical* nature of machine learning (ML) adds complexity to these concepts (Buck et al., 2022). Conventional IT security was primarily concerned with securing databases

Table 2.1 Privacy vs. security in AI systems

Aspect	*Privacy*	*Security*
Definition	Protection of personal/sensitive data	Protection of system/data integrity and availability
Focus	Control over collection, sharing, inference	Resilience to malicious interference
Goals	Confidentiality, anonymity, consent	Integrity, availability, confidentiality
Threat Examples	Membership inference, model inversion, re-identification	Adversarial attacks, poisoning, model extraction, backdoors
techniques	Differential privacy, anonymization, federated learning	Encryption, access control, TEEs, secure deployment
Regulatory role	GDPR, HIPAA, CCPA	NIST AI RMF, cybersecurity standards

or the communication channels; AI security also has to look at how models respond to perturbation or probing. Equally, AI privacy can also be applied to issues such as a model accurately recalling the information about a user (possibly replicating a person's address or face based on training data). In order to overcome these obstacles, frameworks such as National Institute of Standards and Technology (NIST) Privacy Framework and AI Risk Management Framework focus on privacy-by-design (integrating privacy concepts into the systems development process), and trustworthyAI (in which both security and privacy are the main pillars).

2.2.2 Threat models in AI-driven systems

AI-driven systems introduce new *threat models* beyond classical IT security scenarios. Threat modeling in this context means understanding the ways an AI system could be attacked or exploited. Key attack vectors include:

- **Adversarial attacks (evasion):** These are manipulations of a model's *inputs* to trick the AI at inference time. An attacker subtly perturbs input data (images, sensor readings, etc.) so that the model produces an incorrect result, often without the change being noticeable to humans. For example, adding an imperceptible noise pattern to an image might cause a vision system to misclassify a stop sign as a speed limit sign. The field of adversarial ML studies how such attacks exploit the statistical learning nature of ML models (Huang et al., 2011; Sushma et al., 2024). In a *white-box* scenario, the adversary has full knowledge of the model (architecture and parameters) and can compute precise gradients to craft adversarial examples (e.g., using methods like the Fast Gradient Sign Method or Projected Gradient Descent). In a *black-box* scenario, the adversary only has query access to the model's outputs and must either approximate the model or rely on the transferability of adversarial examples (Zhu et al., 2024). White-box attacks tend to be more efficient and potent (since the attacker can optimize directly on the model), whereas black-box attacks require more queries or ingenuity (such as training a surrogate model to mimic the target) (Weng & Wu, 2024). Regardless of the mode, adversarial evasion attacks can pose serious safety risks in AI systems—for instance, causing an autonomous vehicle to ignore a stop sign could lead to accidents. Evaluating a model's robustness to such perturbations (e.g., measuring the fraction of adversarial inputs that cause errors) is now a standard part of AI security testing.

- **Data poisoning:** In poisoning attacks, the adversary tampers with the *training data* of the AI model to undermine its integrity. By inserting carefully crafted malicious samples (or altering existing ones) into the training dataset, the attacker can influence the learned model. This can either degrade the model's overall accuracy (an *availability attack*) or install specific vulnerabilities. A prominent example is the *backdoor attack (Trojan)*, where the model is trained to behave normally on all inputs except those containing a secret "trigger" pattern; when the trigger is present, the model produces a specific attacker-chosen output (Hu et al., 2021). For instance, researchers have shown that it is possible to poison a facial recognition system's training data so that it correctly recognizes all faces *except* when a person wears a particular patterned sticker, at which point the system misidentifies them as a target individual. The trigger can be a visual pattern, a particular phrase in text, etc., and is usually designed to be inconspicuous.

Data poisoning is especially pernicious in scenarios like federated learning (FL) or crowd-sourced data collection, where a malicious contributor might supply tainted data. Modern studies categorize poisoning into *clean-label* (where the malicious samples look ostensibly correct and labels are not obviously wrong) versus *dirty-label* attacks, and *targeted* (aimed at specific outcomes) versus *indiscriminate* (simply degrading performance). Without defenses, poisoning can be hard to detect: the AI model might pass initial validation yet harbor latent malicious behavior. Threat models for poisoning often assume that the attacker has the ability to influence a portion of training data, which is realistic in open systems (e.g., public data sources, or when attackers compromise a data pipeline). Figure 2.1 represents a visualization of common AI attack vectors, including adversarial evasion, data poisoning, clean-label attacks, and backdoor triggers.

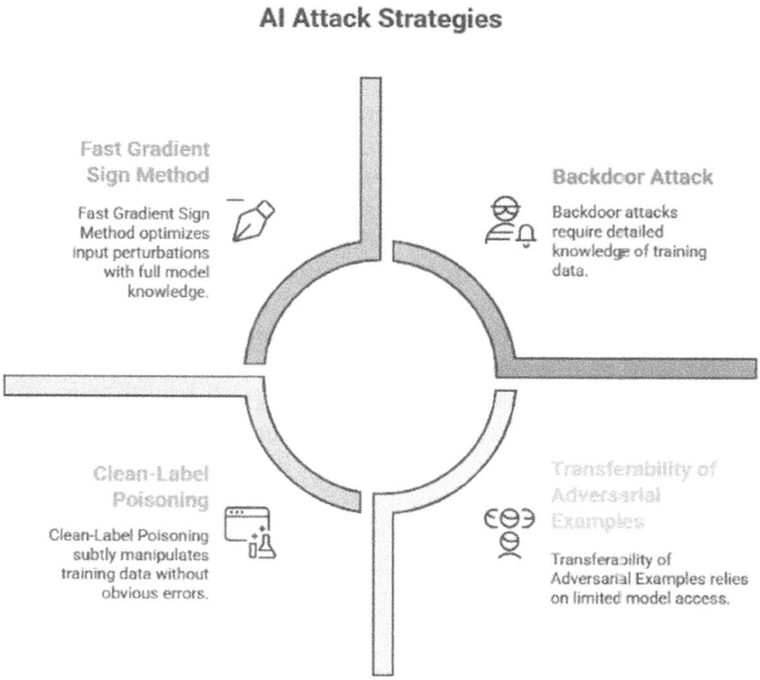

Figure 2.1 Threat model diagram representing complexity of vulnerabilities in AI model.

2.3 DATA PRIVACY IN AI SYSTEMS

2.3.1 Data collection and consent mechanisms

One of the first touchpoints for privacy in AI systems is *data collection*—what data is gathered, from whom, and how. Modern AI, especially in robotics and Internet of Things (IoT), often involves extensive data collection from users or environments, making adherence to privacy principles and regulations critical. Laws such as the European GDPR (General Data Protection Regulation, 2016), the U.S. HIPAA (Health Insurance Portability and Accountability Act, relevant to medical data), and the California CCPA (California Consumer Privacy Act, 2020) set out strict requirements for obtaining user consent and protecting personal data (Jonas et al., 2019; Weng, 2023). Key concepts include *informed consent*, *regulatory compliance, and consent in robotics applications.*

1. **Informed consent:** Users should understand what personal data an AI system or robot is collecting and agree to it. In practice, this might be accomplished via privacy notices, user agreements, or interactive

consent dialogs on a robot's interface. For example, a social robot in a home might ask for permission to record audio to improve its speech recognition. The GDPR elevated the standard for consent—it must be explicit, freely given, specific, and revocable. However, a *gap* has been observed in the robotics industry: many companies do not adequately implement consent mechanisms (Hobel et al., 2014). A 2024 study auditing privacy practices of robotics companies found that over half (55%) of them did *not even disclose which data protection laws they comply with* in their privacy policies. Even more concerning, some robot manufacturers failed to seek *explicit consent* from users before collecting personal information, and in certain cases were found to be selling user data to third parties without clear user awareness. This lack of transparency and consent is "inadvertently or deliberately" turning robots into potential spying devices, as the study title suggests. Such practices are not only ethically problematic but likely illegal under frameworks like GDPR, which can impose hefty fines for non-compliance (Aamot et al., 2013; Fontes et al., 2022).

2. **Regulatory compliance (GDPR, HIPAA, CCPA, etc.):** Regulations provide guidelines and rules that AI systems must follow, especially when handling personal data:

   GDPR (EU): Covers personal data broadly. It mandates principles like data minimization (only collect what is necessary), storage limitation (don't keep data longer than needed), integrity and confidentiality (ensure security), and accountability (organizations must demonstrate compliance). GDPR also grants data subjects rights such as access, rectification, erasure (the "right to be forgotten"), and restriction of processing. For AI, this means users in the EU could request to see what data a system has on them or ask for it to be deleted. If a cognitive robot takes photos in a public space, GDPR would require a legal basis (consent or legitimate interest) and perhaps technical measures like blurring faces of bystanders.

   HIPAA (USA): Relevant in healthcare, it protects identifiable health information. A healthcare robot or AI diagnostic system dealing with patient data must implement safeguards and can only use data for care, payment, or operations unless patient consent is obtained for other uses. Notably, any robot handling patient records would need to follow HIPAA's Security Rule (ensure confidentiality, integrity, and availability of electronic health data) and Privacy Rule (limit uses/disclosures to the minimum necessary). The presence of AI does not change HIPAA's requirements; it just means that the software and hardware running the AI must be within the covered entity's controlled environment, with proper access controls and audit trails (Parker et al., 2020).

CCPA (California): Grants California consumers rights regarding personal data collected by businesses, such as knowing what data is collected, the right to opt-out of its sale, and to request deletion. An AI home assistant sold in California would fall under this law. CCPA's emphasis is on transparency (clear privacy policies) and giving users control (like a "Do not sell my data" option). As mentioned, many robotics companies have not caught up to these expectations—55% did not even state if they comply, meaning users are left in the dark. This is a significant compliance risk and points to a maturity gap in the industry.

Other frameworks include Children's Online Privacy Protection Act (COPPA) (protecting children's data in the United States, relevant if robots interact with kids), Family Educational Rights and Privacy Act (FERPA) (educational records, if AI tutors are used), and emerging laws in various jurisdictions (India's Personal Data Protection Bill [PDPB], Brazil's Lei Geral de Proteção de Dados [LGPD], etc.) (Figure 2.2).

3. **Consent in robotics applications:** Gaining consent in a robotics scenario may require new approaches compared to web/mobile. For instance, if a robot is recording video in a public space, whose consent is needed? Typically, the owner/operator of the robot can consent for their own data, but bystanders might be incidentally recorded. This has led to research on *implicit or contextual consent*—e.g., placing clear signage "Caution: Robot Operating – Video Recording in Progress" in the area to inform people (an analogy to CCTV notices). Socially interactive robots might give a verbal cue: "I'd like to remember your face to recognize you later, is that okay?" and interpret a user's verbal or nod response. Ensuring such consent is logged and honored by the AI (and that it can "forget" data if consent is revoked) is an open technical problem. Some proposed solutions involve maintaining profiles for individuals that include their consent preferences and having the robot's software enforce those (for example, not storing any imagery of people who declined). In healthcare, *informed consent* may involve the robot providing information about data use in human-readable form and obtaining a patient's signature or recorded agreement before proceeding with data-intensive functions.

2.3.2 Privacy-preserving machine learning (PPML)

As AI systems train on and derive insights from data, a key question is: *How can we reap the benefits of machine learning without exposing the underlying sensitive data?* PPML is a field devoted to techniques that enable model training and inference while limiting or preventing the disclosure of

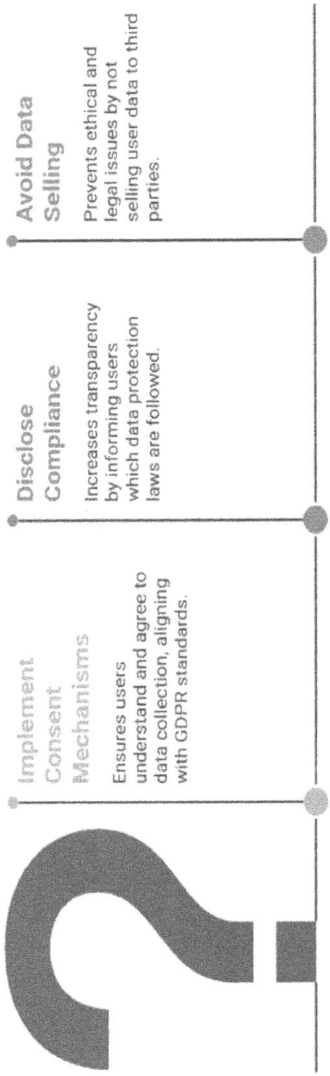

Figure 2.2 Informed consent and regulatory compliance in AI systems.

individual data points. Several leading approaches have emerged in the last decade:

(a) **Federated learning:** Traditional ML involves centralizing all training data on one server or cloud, which poses privacy risks (since raw data leaves the user's device). FL flips the paradigm: models are trained *collaboratively* by distributing the training process to the data sources. For example, in a smartphone application, each phone keeps its user's data locally and trains a local model update; these updates (gradients or model weights) are then sent to a central server which aggregates them (e.g., by averaging) to form a global model (Bourne, 2013; Chen et al., 2024). Crucially, the raw data never leaves the devices. Google popularized this approach for training keyboard prediction models on phones without uploading keystroke logs. Privacy in FL comes from data locality—each participant's data is kept on their side—and from the aggregation process which can be designed to not reveal any single contributor's influence excessively. However, FL has its own challenges and threat models (e.g., a malicious participant could try to infer another's data from the global model, or poison the model, etc.).

(b) **Differential privacy (DP):** DP is a mathematical framework that provides strong privacy guarantees when analyzing or learning from datasets. In simple terms, a mechanism is differentially private if an observer seeing its output (e.g., the trained model or some query answer) can hardly tell whether any single individual's data was included in the input or not. This is typically achieved by adding carefully calibrated random noise to computations. In ML, *Differentially Private SGD* (stochastic gradient descent) is a common technique: during training, one injects noise into the gradient calculations and limits how much influence any single data point can have (often by clipping gradient norms). The result is a model that is slightly less accurate, but provably leaks only a limited amount of information about any specific training example (Finster & Baumgart, 2013). For instance, using DP, a language model could be trained on emails such that it learns general language patterns but is very unlikely to spit out a specific line from someone's private email. DP has been adopted in practice by organizations like Google (for collecting usage statistics with privacy guarantees) and Apple (for crowdsourced learning of typing habits). In robotics or IoT AI, DP could apply when aggregating data from multiple users/environments: e.g., a robot vacuum company could collect maps of users' homes in a differentially private manner, ensuring the aggregated map data doesn't reveal details of any single home. The second application is the publishing of trained models or data sets; DP can provide the sharing of statistical information without exposing the raw data. But it tends to be a matter of balancing the privacy-accuracy trade-off: additional noise (greater privacy) can harm model utility.

(c) **Homomorphic encryption (HE):** HE is the encryption scheme in which the computation can be performed with encrypted data, and the encrypted result will be obtained, and when decrypted, the result will be the same as the result of the operations performed on the plaintext. This is a privacy breakthrough in that it implies that an AI model can be trained on sensitive data without having to view the data in its plaintext. For instance, imagine a cloud-based AI medical diagnosis service: using HE, a hospital could encrypt a patient's medical image, send it to the cloud; the cloud AI (also possibly encrypted or known to the cloud provider) can run the neural network inference on the ciphertext and send back an encrypted prediction that only the hospital can decrypt. At no point does the cloud see the raw image or even the raw prediction (Bourse et al., 2018). There are still performance limitations: deep learning inference via fully homomorphic encryption (FHE) can be orders of magnitude slower than plaintext (though there are reports of continuous improvement, e.g., bootstrapping (recryption) times under a second) (Chialva & Dooms, 2018). But even now, HE is very useful for simpler analytics or where extreme privacy is needed. A related concept is *encrypted model training*, though fully training a large model with FHE is currently not practical due to noise accumulation and complexity. Most uses today are for inference or for training small models or partial steps under encryption (Figure 2.3).

2.3.3 Secure data sharing and storage

Beyond the algorithms of ML, AI systems must handle the practicalities of *storing data securely* and *sharing data with authorized parties in a privacy-protected way*. In many cognitive systems, data flows between devices, cloud servers, and organizations. This subsection covers methods to ensure these data flows and repositories are protected:

Blockchain-based approaches: Blockchain (distributed ledger technology) has been proposed as a tool for secure and transparent data sharing in AI and IoT contexts. The idea is that blockchain's properties—decentralization, immutability, and append-only audit logs—can help address trust issues when multiple parties exchange data or models. For instance, a blockchain can record transactions of a robot sharing sensor data or AI model updates, ensuring that there is a tamper-proof log of who provided data and who accessed it. In a human–robot interaction (HRI) scenario involving sensitive

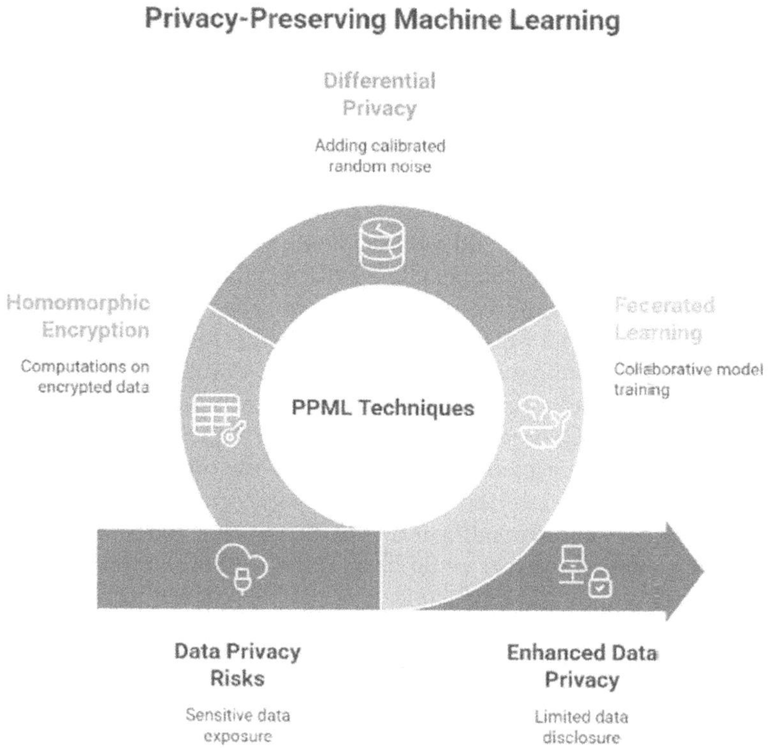

Figure 2.3 Overview of PPML methods.

data (like healthcare robotics across hospitals), a permissioned blockchain could act as a *secure data exchange layer*, where each data upload/download is recorded and only authorized nodes (hospitals, researchers) can participate. In industrial robotics, using blockchain for logging robot sensor readings and commands can provide an audit trail that attackers cannot easily alter, which is useful for forensic analysis after an incident. However, blockchain is not a silver bullet: performance and scalability can be issues (though permissioned blockchains improve on public ones), and privacy on a ledger can be tricky (since by default all nodes see all transactions). Techniques like zero-knowledge proofs and encryption on blockchain are being developed to address that, so one can prove certain data or computations exist without revealing the data itself. The key takeaway is that blockchain can improve *data accountability and integrity* in multi-agent AI systems by ensuring all parties have a consistent view of the data history and that records cannot be retroactively changed (Dhieb et al., 2020; Elkhodr et al., 2024).

Secure multiparty computation (SMC): SMC allows a set of parties to jointly compute a function over their inputs while keeping those inputs private from each other. Unlike HE, which often assumes a semi-trusted server, SMC typically involves interactive protocols among parties who each hold a piece of the input. In AI data sharing, SMC can enable *collective data analysis or model training* where, say, five hospitals compute a ML model on the union of their datasets without any hospital seeing data from the others. This is done through exchanging encrypted or secret-shared values and making sure that at the end only the final model (or whatever output is desired) is revealed. SMC can be achieved by techniques like secret sharing (each party holds a random share of every data value, and only together can they reconstruct the true values) and garbled circuits for boolean operations. For example, two companies want to use their combined customer data to train a recommendation system but are competitors and cannot share databases. With SMC, they can train a model as if their data were pooled, but each only sees encrypted or random bits of the other's data during the process, and perhaps only the final model parameters (which themselves might be considered not revealing individual data). The outcome is that they get a better model than either could have alone, without violating customer privacy (Junaid et al., 2022; Mustafa et al., 2024; Salah et al., 2019).

Data anonymization and pseudonymization: These are more traditional but still important techniques, especially as mandated by laws. *Anonymization* involves removing or obscuring personal identifiers from data such that individuals are not identifiable. *Pseudonymization* replaces real identifiers (like names) with pseudonyms (like random IDs) so that data can still be linked across datasets via the pseudonym but without revealing the original identity. In AI contexts, anonymization might be applied to training data *before* feeding it into the model. For example, a dataset of user messages could be stripped of names, email addresses, and phone numbers, or a dataset of images could have faces blurred if the identity is not relevant for the task. Many organizations have policies to anonymize data after collection and before analysis to reduce privacy risk. A major challenge is that AI thrives on rich data, and anonymization often involves removing information (which could be useful for the model). For instance, removing exact geolocation or date of a transaction to anonymize may make fraud detection less accurate. This trade-off can sometimes be mitigated by clever pseudonymization—e.g., replacing actual location with a region or transforming dates to age or relative time (Ahakonye et al., 2024; Ebrahim et al., 2022). Pseudonymization is useful in longitudinal studies or multi-source data integration, where you need to track a user or entity's data across time or datasets without using their real identity. A concrete example in robotics: suppose robots in a mall are capturing footage to analyze foot traffic patterns. To protect privacy, each person detected could be assigned a random ID (pseudonym) for that day. The analysis can then track that "ID123" moved from store A to B to C, without ever knowing who ID123 really was. The raw video might be discarded, and only those pseudonymized trajectories kept for analysis.

2.4 SECURITY CHALLENGES IN AI MODELS

2.4.1 Vulnerabilities of ML models

AI models, particularly those based on ML and neural networks, have introduced new *vulnerabilities* that differ from traditional software vulnerabilities. Two notable categories of such vulnerabilities are *model extraction* (covered in threat models) and *backdoor attacks*, but there are others like model misuse or logic flaws. Here, we highlight model extraction and backdoors, as well as the general nature of ML model weaknesses:

Model extraction (stealing) revisited: ML models may also be a property, which is trained using valuable data. Model extraction attacks enable an adversary to perform a copy of a model by making queries and monitoring responses. This is a weakness since it undermines the privacy of the model parameters (and, indirectly, the privacy of the training data [a stolen model can incorporate features of the training data]) by exposing the model to systems other than the original one. This compares to reverse engineering an application programming interface (API) in cloud ML services. The vulnerability lies in the fact that most ML models will be happy to provide highly informative responses (e.g., class probabilities) to just about any input. Attackers can use this to conduct what can be referred to as a kind of system identification. Experiments have demonstrated that a large access to queries can be used to approximate even complicated deep-learned models (such as image classifiers or language models) with a minor accuracy loss (Sangeetha et al., 2024; Zhang et al., 2020). Such substitute models can be used to mount further attacks (like generating adversarial examples offline). They can also violate intellectual property—imagine an attacker replicating a paid ML service and offering it for free or using it in unapproved ways. From a security standpoint, mitigating model stealing might involve *limiting query rates*, *reducing output detail*, or *using watermarking*. Watermarking embeds a hidden signature in model behavior to trace stolen copies (for instance, the model might respond in a specific quirky way to a set of trigger inputs known only to the owner, allowing detection of a stolen model if it replicates that behavior). However, these defenses are not foolproof and often trade-off with utility. It is a growing concern, especially as AI models become larger (e.g., large language models) and more expensive to train—there is a financial incentive for adversaries to steal rather than train from scratch.

Backdoor and Trojan attacks: A *backdoor attack* on an AI model was described earlier as a kind of poisoning where a trigger in inputs causes a specific malicious behavior. From the model's perspective, a backdoor is an unintended function embedded into it (Dasgupta & Collins, 2019). For instance, a traffic sign classifier might be trained (poisoned) so that images of a stop sign with a small yellow sticker are misclassified as speed limit 45. Under normal circumstances (no sticker), it works fine; with the sticker, it consistently gives the attacker's desired wrong output. This is a severe vulnerability because it may not be discovered in regular testing—if the trigger is rare or looks innocuous, the model owner might never test that exact combination. Attackers could exploit this at deployment (e.g., by physically placing that yellow sticker on a real stop sign to fool autonomous cars). Backdoors can enter models via training data poisoning (as described) or even during model fine-tuning (imagine downloading a pre-trained model from an untrusted source that has a backdoor in it). A related concept is *model Trojaning*: where the attacker directly modifies the model's parameters (if they have that access) to insert a backdoor, akin to how a software Trojan might add malicious code. According to a survey, backdoor attacks on deep learning have been demonstrated in various domains (vision, natural language processing [NLP], audio) and can be quite stealthy.

Model misuse and evasion of guardrails: The vulnerability that is less widely discussed is that AI models may be abused in situations they were not intended to be applied to, and it will result in security problems. As an example, a generative language model (such as GPT) can be prompted in such a manner that it generates non-permitted content or even malicious code (such as prompt injection attacks on generative AI). Although not a security flaw in the code sense of the word, it is a weakness of the model in its correspondence to intended use. Bypassing filters: A content-filtering AI may be fooled by adversarially generated text that perplexes the filter. Or an image recognition system might be *overly confident* in its output, which can be exploited in social engineering (imagine a facial recognition system that occasionally misidentifies someone, and an attacker uses that to impersonate someone in a physical access system). These issues show that ML models often lack *graceful failure modes*; when given inputs outside their trained distribution, they can behave unpredictably. This is a vulnerability if an attacker can deliberately feed such inputs. Unlike traditional software, where unusual input might cause a crash (which is itself a problem), in ML, unusual input might cause a *highly confident wrong action*, which could be worse (the system doesn't even realize that it is uncertain). Adversaries exploit this in evasion attacks and also in causing models to output sensitive data (if a model is trained on private data, a clever input might get it to spout out that data—a form of model inversion as discussed).

In general, ML models are vulnerable because they are *data-defined* programs rather than logic-defined. They interpolate and sometimes extrapolate from training data, which adversaries can take advantage of by either manipulating the training (poisoning/backdoors) or manipulating the input (adversarial examples) or the output (extraction). Moreover, many ML models are overparameterized (more parameters than data points in many deep nets), making them prone to memorization and unforeseen behaviors. They also often operate as black boxes even to their creators, which complicates debugging and security analysis. Traditional software security has practices like code audits, formal verification of critical components, etc.—analogous practices for ML (like formally verifying certain robustness properties) are nascent (some progress in verifying simple neural networks' robustness to adversarial input).

2.4.2 Secure deployment of AI in cognitive machines

Attaining the security of AI models is not merely a matter of the algorithms and training, but also a matter of the deployment. The cognitive machines may operate on the edge (where the device, robot, or local gateway executes it) or on the cloud (where it transmits data to servers to be processed). Every paradigm possesses security and privacy implications, and at the hardware level:

Edge vs. cloud security: *Edge AI* refers to running AI computations on local devices (robots, IoT sensors, smartphones) rather than sending data to a centralized server. From a privacy standpoint, edge AI is often beneficial: data can be processed where it is collected, without transmitting sensitive raw data over networks. For instance, an AI-driven security camera that does face recognition on-device can decide whether to alert or not without streaming video to the cloud (thus reducing exposure of those video feeds). This local processing inherently "enhances security and reduces the risk of widespread exposure" since there is no single cloud repository of all users' data that could be breached. Additionally, edge computing can operate even offline or with limited connectivity, which is important for reliability and safety in critical applications (e.g., a self-driving car must make decisions immediately, without waiting for cloud responses). However, edge devices come with constraints: they have limited computational power, and implementing strong security (like full disk encryption, secure boot, frequent patches) can be more challenging, especially for low-cost IoT/robotics hardware. Moreover, physical access to edge devices is often easier for attackers. A robot or sensor in the field could be stolen or tampered with directly, which is less of an issue for a cloud server in a data center. This means *physical security* (tamper-resistant casings, etc.) and techniques like secure boot (to prevent loading altered firmware) are crucial for edge AI. In comparison, Cloud AI concentrates the heavy calculation. Cloud providers generally have advanced cybersecurity services, sensors, and infrastructure security (firewalls, distributed denial of service [DDoS], etc.). Cloud security practices can offer adequate protection to data stored on the cloud, and cloud systems can scale up computational defenses (such as running anomaly detectors on user input across numerous instances). The disadvantage is that as the data moves out of the edge to the cloud, it moves along networks where it can be intercepted (unless encrypted) and is stored in the servers (which can be desirable targets in breaches). Cloud use can also be complicated by compliance with the laws governing data residency (not all sensitive data can be transferred out of some jurisdictions). Practically, most systems adopt a hybrid model: a little pre-processing on the edge, and more intensive analytics in cloud, in an attempt to have the best of both worlds. For example, an autonomous vehicle might do immediate obstacle detection on edge but send aggregated driving data to cloud for fleet learning. Security architecture must then cover both fronts: encrypt data in transit between edge and cloud (e.g., transport layer security [TLS]), authenticate devices to the cloud to prevent impersonation, and ensure cloud APIs are protected from abuse. There is also the aspect of *update security*: edge devices need firmware and model updates. Ensuring these updates come from legitimate sources and are applied correctly is vital (compromised update mechanism can inject malware or tainted models). Cloud-hosted models can be updated more easily, but ensuring that edge devices trust only authentic updates involves cryptographic signing and verification (Kumar et al., 2023).

Hardware-level vulnerabilities (side-channel attacks, etc.): Leakage or attack may also be a source of the hardware executing AI. Side-channel attacks use physical emanations or common resources to deduce secrets. In case of AI, this entails such things as: monitoring the power usage or electromagnetic emittances of a device during a neural network run to deduce information about either the model or the input data (Cammarota et al., 2020). For example, a research paper titled "I Know What You See" demonstrated that by analyzing the power usage of a hardware accelerator for a CNN, one could deduce which image (from a set of known images) the network was processing. Essentially, the pattern of power usage was unique to different inputs. This is a scary prospect: even if the AI computation is on a device in a secure environment, an attacker with a power meter or maybe even through the power outlet could spy on what the AI is seeing or deciding. Similarly, *cache side-channels* have been used in scenarios like Software Guard Extension (SGX) enclaves to extract neural network model parameters (by triggering cache accesses and timing them). These attacks require some sophistication, but they are real—especially if the attacker can run some code on the same device (like a co-located malicious app that monitors resources). Another hardware threat is *fault injection*—attackers might use techniques like voltage glitching or laser fault injection to make a device's processor miscompute in a way useful to them (e.g., skip an authentication check or cause a memory read to bypass bounds) (Dash, 2024).

Trusted execution environments (TEEs) like Intel SGX and ARM TrustZone are a partial solution to hardware threats by isolating sensitive code and data from the rest of the system, even if the OS is compromised. Using TEEs, one can load a decryption key or an ML model into an enclave such that even side-channel ability is limited (SGX enclaves still had some side-channel issues historically, but improvements are underway). TEEs can ensure that even if an attacker compromises the main system, the AI model's weights and computations in the enclave remain confidential and unaltered. For example, a cloud server with SGX can run a deep learning inference on sensitive data inside an enclave so that the cloud operator cannot see the data or model directly (Browne et al., 2022).

2.5 PRIVACY AND SECURITY IN COGNITIVE ROBOTICS

2.5.1 HRI and data sensitivity

HRI lies at the heart of cognitive machines operating in social or collaborative settings. As robots enter homes, workplaces, and public spaces, they invariably handle data that is personal or sensitive. Two major concerns in this context are the *sensitivity of the data collected in HRI* (biometric, behavioral, contextual information about people) and the *ethical and trust implications* of how robots handle that data.

Sensitive data in HRI: Humans are sensed in multiple modalities by robots, including cameras that see faces and expressions, microphones that record voices, touch or proximity sensors to note presence and movement, etc. (in medical robots, other types of specialized sensors may include heart rate or other biometrics). This information is full of personal information. An example: A social robot camera is capable of recognizing people (face recognition can determine identity, gender, approximate age, emotional state, etc.). Although not face recognition, an act of merely capturing the activity in the home of a person is very sensitive (finding out the routine activities, habits, even medical conditions or personal relations). A microphone on a robot companion can capture confidential chats or the voice tone of a person (which can be used to gauge mood or stress). Voice information can be as identifying as a fingerprint—it is possible to identify them by the timbre of their voice, and it is possible to milk speech information of all kinds of personal information. In caring robots, including eldercare robots, the robot may deal with health-related information: medication times, vital signs, or the last time a person walked (which may signify a fall or a medical condition). These would be safeguarded health information under laws such as the HIPAA. Behavioral data, such as the manner in which an individual treats the robot (frequency, commands issued, emotional responses), may result in a profile of his preferences and emotional condition. An example is where a child has an educational robot tutoring them that implicitly acquires information on the learning disabilities or emotional condition of the child through sensor inputs (Pierce, 2020).

Ethical and trust implications: Privacy is sensitive to the manner robots manage it, which interferes with human confidence and embracement. When individuals begin thinking that my robot vacuum could be spying on them and sending them images of the living room to the manufacturer (it is no baseless fear, as there are certain problems with the IoT lately), they might not embrace the technology. There are also some ethical dilemmas: to what extent should a robot reveal itself to what it sees? Take the example of a healthcare robot that observes symptoms of depression in a patient: does it have the ethical responsibility to divulge that to a caregiver (probably violating the privacy of the patient, but in their best interest)? They are complicated problems that cannot be fixed in a single size. Transparency is generally useful in building trust: robots should tell how they see and use their data. There are guidelines that propose that the robots can be equipped with icons or indicator lights on when to record (Mahmoud et al., 2023).

Human behavior changes: There is evidence that the mere presence of recording-capable technology changes behavior (people might self-censor or avoid the robot if concerned). If robots become widespread, will people feel they must always be "on guard" as if in public? That would erode the very comfort of private life. Thus, maintaining *clear boundaries* for when/where robots record and ensuring that in private spaces the control stays with the human is crucial to avoid that chilling effect (Figure 2.4).

2.5.2 Privacy-aware robotic perception

Robotic perception refers to the processes by which robots interpret sensor data (vision, audio, light detection and ranging [LIDAR], etc.) to understand their environment. Making this perception *privacy-aware* means designing perception systems that minimize the capture or exposure of sensitive information, and *secure* means protecting the sensor data from interception or tampering. We explore techniques and challenges for vision and speech processing in robots, and securing sensor data generally:

Secure vision processing: Cameras are one of the most powerful sensors for robots, but also the most privacy-intrusive. A camera can capture faces, documents, computer screens, etc., often incidentally. To reconcile

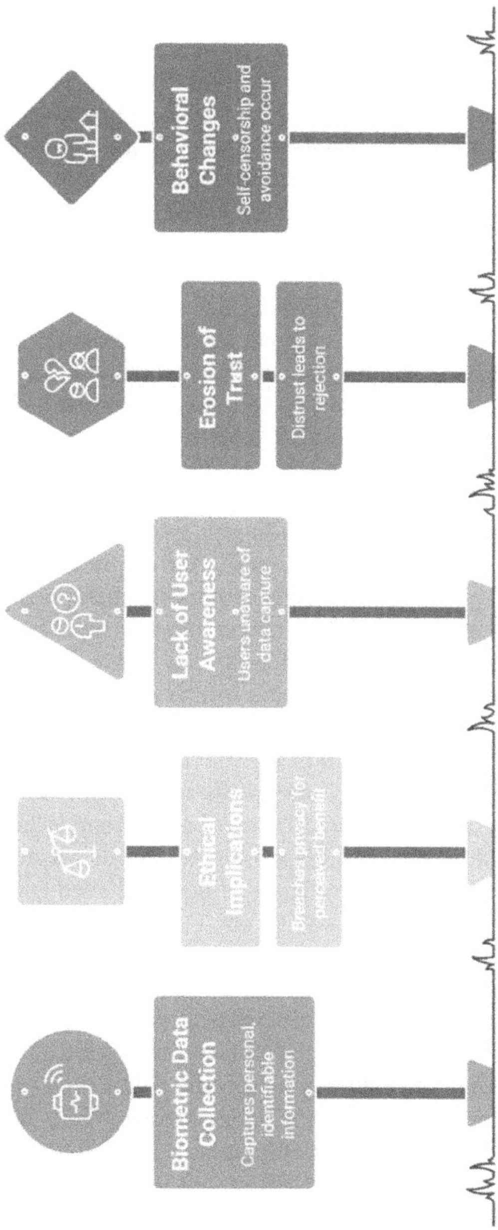

Figure 2.4 HRI data sensitivity.

the utility of vision with privacy: One strategy is using *on-device computer vision* to extract only the needed information from images and immediately discarding or sanitizing the raw images. For example, if a home robot needs to navigate around people, it might only need to detect human shapes and locations, not identify who they are. In that case, the robot's vision algorithm can be tailored to output "anonymous" data, like bounding boxes of humans without facial details. There are even approaches to design neural networks that do not require reconstructable images at intermediate layers, thereby inherently limiting what could be stolen from them (Pittaluga & Koppal, 2017). **Blurring or filtering:** If a robot does need to transmit or save an image (say, to get remote assistance or for training improvements), it can first blur faces or remove identifiable features. Some robots use cartoonified representations for remote monitoring—for instance, representing people as simple figures or blobs on a map rather than detailed video. An example is certain telepresence robots that deliberately lower video resolution or replace the background to avoid revealing environmental details. **Focus on task-relevant features:** A robot might have multiple computer vision models, some for detecting objects, some for recognizing people. When the application does not need person identification, such sections of the model may be disabled or omitted. Correspondingly, a segmentation mask might be applied to block out parts of an image that are not required in the task (such as concealing any text on walls or papers when purely doing obstacle detection). Secure vision is also the prevention of outsiders from accessing the video feed. Any footage or video streams stored should be encrypted strongly (with keys available only to authorized users or the robot owner). Also, authentication of any connection to video output (to eliminate spoofing or viewing unauthorised). In addition, the robot is expected to process by default privacy regulations any personally identifiable information (PII) in pictures. As an example, when it is used to recognize (approved) family members, then it should preferably store face data in a secure representation (e.g., not raw images but feature embeddings, and even encrypted or stored locally).

Privacy in speech processing: Audio is delicate because humans might utter passwords, health data, etc., in the presence of a robot. Certain privacy-conscious speech processing principles: Use on-device wake-word detection. This makes sure that the microphone on the robot only hears or processes significant audio when a particular trigger word ("Hey Robot") has been uttered, and otherwise it ignores ambient conversation (this is the strategy used by smart speakers such as Alexa, although people have expressed worries over false triggers that might cause unintentional recording). By not streaming or storing audio until awakened, incidental overhearing is reduced. When speech is processed (like converting speech to text), prefer on-device processing if feasible, especially for local commands ("turn on the light") to avoid sending potentially sensitive voice data to the cloud. If cloud processing (for more complex requests) is needed, send it over encrypted channels and immediately delete after use on the server side (with transparency in privacy policy). If a robot records conversations (say for telepresence or elder monitoring for safety reasons), this should be explicitly consented to and indicated (maybe a light that turns on during recording), and recordings should have restricted access (Tang et al., 2022). For example, if family members use a robot to check on an elderly relative via audio, those audio streams should be accessible only to them through secure accounts. **Speaker identification:** If the robot identifies people by voice (biometric identification), that voiceprint is sensitive personal data (like a fingerprint). It should be treated like a password—stored in a hashed form if possible, or at least encrypted. The system should also allow opting out of voice ID (maybe use QR codes or other means for ID if voice storage is an issue for a user). **Noise addition for privacy:** In some research, adding a subtle noise or encryption-like transformation to audio can allow a service to do needed processing but not clearly recover the spoken content. These are not widely deployed, but an idea, e.g., if sending to cloud, distort the audio in a reversible way (key held by user) so that even if intercepted, it is garbled.

Sensor data protection: Beyond vision and audio, robots have other sensors (LIDAR, sonar, infrared, etc.). These might seem less privacy-invasive (LIDAR gives 3D point clouds with no color or texture). However, even point clouds can reveal a lot (shapes of furniture, presence of humans as blobs). For industrial sensors (temperature, pressure), it might be more about security than personal privacy (ensuring those readings are not falsified by an attacker to cause damage). A general principle is *data minimization at source*: Only collect the range and resolution needed (Yu et al., 2022). For instance, a domestic robot's LIDAR might not need millimeter accuracy of everything, just enough to avoid obstacles—so maybe it can aggregate points into a lower-resolution grid that reveals less fine detail of the environment. **Encrypting sensor streams:** If sensors are external or networked (like a separate CCTV that feeds into the robot's system), use secure links (so no one can snatch the feed via Wi-Fi interception). **Authentication:** Ensure that each sensor's data is coming from a legitimate source (to avoid sensor spoofing, where an attacker injects fake data). This can be done with secure hardware on sensors that sign data, or by cross-checking between multiple sensor modalities (if camera sees nothing but LIDAR sees an object, one might be spoofed). **Storage:** Many robots log sensor data (for debugging, ML improvements, or regulatory reasons). Those logs should be protected. A notable example: some robot vacuums have cameras, and images from them were leaked or shared, causing user concern. Companies should either avoid storing raw sensor feeds or highly restrict and anonymize them. If logs are needed (e.g., for accident investigation in self-driving cars), access to them must be carefully controlled and possibly data scrubbed for privacy (faces blurred, license plates, etc. in camera logs).

2.6 CHALLENGES AND FUTURE DIRECTIONS

Despite significant progress in securing AI systems and preserving privacy, many challenges remain unresolved. This concluding section highlights some issues and future directions for research and practice, as we look toward increasingly cognitive and autonomous machines in society.

1. **Scalability of privacy-preserving AI methods:** Techniques like DP, SMC, and HE have shown they can protect data, but scaling them to the massive models (billions of parameters) and datasets of today is difficult. For instance, differentially private training often incurs a major utility cost for complex models, and HE that can handle a deep neural network is extremely computationally intensive. As AI models continue to grow (think of future GPT-5 with trillions of parameters, or national-scale robotics networks), making privacy techniques efficient enough is a challenge. Research is needed on more clever algorithms (perhaps exploiting structure in data or models) to reduce the performance overhead. The future might also involve specialized hardware to accelerate privacy-preserving computations (there are startups and academic projects on Application-Specific Integrated Circuits [ASICs] for HE, etc.). Until these methods become more practical, organizations will sometimes skip them and rely on policy or trust, which is a risk. So advancing the state-of-the-art in PPML such that it can be routinely applied to large-scale systems is crucial to truly bake privacy into all AI systems.
2. **Balancing security, accuracy, and efficiency:** Often, enhancing security (like adversarial robustness) or privacy (like DP) comes at the cost of model accuracy or system efficiency. For example, an adversarially trained model might be more secure but slightly less accurate on benign inputs, or require more computation. Similarly, encrypting all sensor data might add latency that an autonomous system cannot afford. The challenge is finding sweet spots or new techniques that provide security without such heavy trade-offs. There's promising work on *certified robustness* where models can be proven to resist certain attacks within a bound, but often those models are smaller or simpler.
3. **Interdisciplinary gaps:** Privacy and security in AI cannot be solved by technologists alone, nor by policymakers alone. There are interdisciplinary gaps in understanding that need bridging:
   (a) Law and AI: Many regulators/lawyers are not deeply familiar with how AI works (the technology is complex and evolving).

Conversely, many AI developers are not well-versed in legal obligations or ethical theory. The future calls for more professionals who have dual literacy (perhaps specialized AI compliance officers, or more collaboration in drafting regulations that are technically informed). Similarly, ethical frameworks need to be translated into engineering requirements in a concrete way (an ongoing challenge).

(b) Human factors: Security systems often fail because they do not account for human behavior. If privacy/security features make an AI product too inconvenient, users might disable them (like covering robot cameras, etc.). So human–computer interaction (HCI) research needs to ensure that security/privacy features are user-friendly and intuitive. Building user trust interfaces (like effective indicators of AI recording status, easily accessible privacy settings) is an open area for HCI.

(c) Socio-economic context: The impact of AI on privacy/security may disproportionately affect certain groups (e.g., surveillance AI is often more deployed in marginalized communities). More research and governance should involve social scientists to ensure AI deployments do not exacerbate inequalities or create "surveillance underclass." This ties into fairness and civil liberties beyond the technical realm.

2.6.1 Future directions for cognitive AI systems

(a) **Adaptive security:** Future robots might learn and adapt their policies online. This raises the question: how do we verify and secure systems that evolve after deployment? Traditional static verification might not hold if the model changes. There is work on "lifelong learning under constraints," ensuring that as an AI adapts, it does not drift into unsafe or non-compliant behaviors. This will be crucial for long-lived cognitive machines.

(b) **Explainability vs. complexity:** With more cognitive ability, systems might become even more opaque (e.g., an autonomous system with an inner world model making complex decisions). Ensuring they remain explainable and thus auditable is an open challenge that will persist and increase in importance.

(c) **Edge AI and federated robotics:** Computation is increasingly being pushed to the edge (on privacy and latency) as we described. It is possible that in the future, swarms of robots will be able to learn (FL) or even make group decisions. Security in distributed decision-making (to ensure one rogue robot does not destroy the group) and privacy (to ensure the personal data of one robot is not disclosed during collaboration) are further generalizations of our multi-agent problems at present. This will require new protocols, perhaps building on technologies such as blockchain, or new consensus mechanisms resilient to bad actors.

(d) **AI governance evolution:** We expect a proliferation of *auditing frameworks* and possibly *certifications for AI systems* (similar to how safety-critical systems get certified). Developing methodologies for AI audits (including privacy and security testing) is an open area. Some envision something like an "FDA for Algorithms" or mandatory third-party audit regimes for high-risk AI. Ensuring these governance solutions are effective without stifling innovation is tricky, and will be a focus of near-future policy and research. Pilot projects (like algorithm registries, incident databases) might become standard tools.

(e) **Public engagement and trust:** As cognitive machines become more visible in daily life, maintaining public trust will be an ongoing effort. Open challenges include how to *educate the public* about AI systems (to prevent both undue fear and undue trust), and how to incorporate public values into design (maybe via participatory design or citizen panels influencing AI guidelines for their community). Social acceptance can be as much a barrier as technical issues; we have seen some cities ban certain AI tech due to public pushback. Future AI deployment will need more proactive trust-building measures beyond just meeting minimum compliance.

2.7 CONCLUSION

In this chapter, we have discussed the complex interplay of privacy, data security, and AI systems, especially in regards to cognitive machines and robotics. When used without significant precautions, AI–robotics can cause the violation of individual privacy, damage user trust, be exploited maliciously, and even cause bodily injury. In contrast, privacy and security designed in and governed become sources of innovation—users and society will much more readily adopt cognitive machines that show evidence of respecting their rights and safeguarding them. The realization of privacy and data security in AI is a constant process. As AI and robotics develop, new threats will always be developed to be countered, and new ethical issues will always be addressed. Nevertheless, the main conclusion of our analysis is that an active and principled stance on privacy and security is not going to hinder AI development, but is, on the contrary, a facilitator of it. Cognitive machines can become what they promise to be, namely, augmenting human abilities and making the world a better place through platforms that build trust among users and society through robust security and ethical governance.

REFERENCES

Aamot, H., Kohl, C., Richter, D., & Knaup, P. (2013). Pseudonymization of Patient Identifiers for Translational Research. *BMC Medical Informatics and Decision Making*, *13*(1), 75. https://doi.org/10.1186/1472-6947-13-75

Ahakonye, L. A. C., Nwakanma, C. I., & Kim, D. S. (2024). Tides of Blockchain in IoT Cybersecurity. *Sensors*, *24*(10), 3111. https://doi.org/10.3390/s24103111

Bourne, I. (2013). Personal, Pseudonymous, and Anonymous Data: The Problem of Identification. *XRDS Crossroads the ACM Magazine for Students*, *20*(1), 27–31. https://doi.org/10.1145/2508972

Bourse, F., Minelli, M., Minihold, M., & Paillier, P. (2018). Fast Homomorphic Evaluation of Deep Discretized Neural Networks, pp. 483–512. https://doi.org/10.1007/978-3-319-96878-0_17

Browne, J. T., Bakker, S., Yu, B., Lloyd, P., & Allouch, S. B. (2022). Trust in Clinical AI: Expanding the Unit of Analysis. *Frontiers in Artificial Intelligence and Applications, 354*, 93–113. https://doi.org/10.3233/faia220192

Buck, C., Doctor, E., Hennrich, J., Jöhnk, J., & Eymann, T. (2022). General Practitioners' Attitudes toward Artificial Intelligence–Enabled Systems: Interview Study. *Journal of Medical Internet Research*, *24*(1), e28916. https://doi.org/10.2196/28916

Cammarota, R., Schunter, M., Anand, R., Boemer, F., Kiss, A. I., Treiber, A., Weinert, C., Schneider, T., Stapf, E., Sadeghi, A. R., Demmler, D., Stock, J., Chen, H., Hussain, S. U., Riazi, S., Koushanfar, F., Gupta, S., Rosing, T. S., Chaudhuri, K., . . . Norton, P. A. (2020). Trustworthy AI Inference Systems: An Industry Research View. arXiv preprint arXiv:2008.04449, 2020. Accessed: Aug. 10, 2020. [Online]. Available: https://arxiv.org/abs/2008.04449. https://doi.org/10.48550/arxiv.2008.04449

Chen, J., Yan, H., Liu, Z., Zhang, M., Xiong, H., & Yu, S. (2024). When Federated Learning Meets Privacy-Preserving Computation. *ACM Computing Surveys*, *56*, 1–36. https://doi.org/10.1145/3679013

Chialva, D., & Dooms, A. (2018). Conditionals in Homomorphic Encryption and Machine Learning Applications. *Queen's University Computer Science Technical Report No. 2018-636*, 2018. [Online]. Available: https://research.cs.queensu.ca/TechReports/Reports/2018-636.pdf. https://doi.org/10.48550/arxiv.1810.12380

Dasgupta, P., & Collins, J. B. (2019). A Survey of Game Theoretic Approaches for Adversarial Machine Learning in Cybersecurity Tasks. *AI Magazine*, *40*(2), 31–43. https://doi.org/10.1609/aimag.v40i2.2847

Dash, B. (2024). Zero-Trust Architecture (ZTA): Designing an AI-Powered Cloud Security Framework for LLMs' *Black Box Problems*, *4*(2), 1–5. https://doi.org/10.54026/ctes/1058

Dhieb, N., Ghazzai, H., Besbes, H., & Massoud, Y. (2020). Scalable and Secure Architecture for Distributed IoT Systems. 2020 IEEE Technology & Engineering Management Conference (TEMSCON). https://doi.org/10.1109/temscon47658.2020.9140108

Ebrahim, M., Hafid, A., & Elie, E. (2022). Blockchain as Privacy and Security Solution for Smart Environments: A Survey. *arXiv preprint arXiv:2203.08901* [cs.CR],

Mar. 2022. [Online]. Available: https://arxiv.org/abs/2203.08901. https://doi.org/10.48550/arxiv.2203.08901

Eleshin, F., Iradukunda, P., Ruberamitwe, D. I., & Ishimwe, E. (2024). Your Robot Might Be Inadvertently or Deliberately Spying on You: A Critical Analysis of Privacy Practices in the Robotics Industry. In *ICT4Ag 2024: Proceedings of the 11th International Conference on ICT for Agriculture*, Lisbon, Portugal, 2024, pp. 1-12. [Online]. Available: https://www.scitepress.org/Papers/2024/124222/124222.pdf

Elkhodr, M., Khan, S., & Gide, E. (2024). A Novel Semantic IoT Middleware for Secure Data Management: Blockchain and AI-Driven Context Awareness. *Future Internet*, *16*(1), 22. https://doi.org/10.3390/fi16010022

Feretzakis, G., Papaspyridis, K., Gkoulalas-Divanis, A., & Verykios, V. S. (2024). Privacy-Preserving Techniques in Generative AI and Large Language Models: A Narrative Review. *Information*, *15*(11), 697. https://doi.org/10.3390/info15110697

Finster, S., & Baumgart, I. (2013). Pseudonymous Smart Metering without a Trusted Third Party. 2013 12th IEEE International Conference on Trust, Security and Privacy in Computing and Communications, pp. 1723–1728. https://doi.org/10.1109/trustcom.2013.234

Fontes, M. M., Pedrosa, D., Morgado, L., & Cravino, J. (2022). Anonymizing Student Team Data of Online Collaborative Learning in Slack. 2022 International Conference on Advanced Learning Technologies (ICALT), pp. 11–12. https://doi.org/10.1109/icalt55010.2022.00009

Hobel, H., Schrittwieser, S., Kieseberg, P., & Weippl, E. (2014). Anonymity and Pseudonymity in Data-Driven Science, pp. 124–130. https://doi.org/10.4018/978-1-4666-5202-6.ch013

Hu, Y., Kuang, W., Qin, Z., Li, K., Zhang, J., Gao, Y., Li, W., & Li, K. (2021). Artificial Intelligence Security: Threats and Countermeasures. *ACM Computing Surveys*, *55*(1), 1–36. https://doi.org/10.1145/3487890

Huang, L., Joseph, A. D., Nelson, B., Rubinstein, B. I. P., & Tygar, J. D. (2011). Adversarial Machine Learning. In *Proc. 4th ACM Workshop on Artificial Intelligence and Security (AISEC 2011)*, October 2011, pp. 43–58. doi: 10.1145/2046684.2046692

Jonas, S., Siewert, S., & Spreckelsen, C. (2019). Privacy-Preserving Record Grouping and Consent Management Based on a Public-Private Key Signature Scheme: Theoretical Analysis and Feasibility Study. *Journal of Medical Internet Research*, *21*(4), e12300. https://doi.org/10.2196/12300

Junaid, S. B., Imam, A. A., Balogun, A. O., Silva, L. C. D., Surakat, Y. A., Kumar, G., Abdulkarim, M., Shuaibu, A. N., Aliyu, G., Sahalu, Y., Abdullahi, M., Mohammed, T. Y., Abdulkadir, B. A., Abba, A. A., Kakumi, N. A. I., & Mahamad, S. (2022). Recent Advancements in Emerging Technologies for Healthcare Management Systems: A Survey. *Healthcare*, *10*(10), 1940. https://doi.org/10.3390/healthcare10101940

Kumar, A., Krishnamoorthy, B., & Bhattacharyya, S. S. (2023). Machine Learning and Artificial Intelligence-Induced Technostress in Organizations: A Study on Automation-Augmentation Paradox with Socio-Technical Systems as Coping Mechanisms. *International Journal of Organizational Analysis*, *32*(4), 681–701. https://doi.org/10.1108/ijoa-01-2023-3581

Liao, T., Li, T., & Nadkarni, P. (2025). A Survey on Secure Machine Learning. *arXiv preprint arXiv:2505.15124*.
Liu, B., Ding, M., Shaham, S., Rahayu, W., Farokhi, F., & Lin, Z. (2020). When Machine Learning Meets Privacy. *ACM Computing Surveys (CSUR)*, *54*, 1–36. https://doi.org/10.1145/3436755
Liu, X., Xie, L., Wang, Y., Zou, J., Xiong, J., Ying, Z., & Vasilakos, A. (2021). Privacy and Security Issues in Deep Learning: A Survey. *Ieee Access*, *9*, 4566–4593. https://doi.org/10.1109/ACCESS.2020.3045078
Mahmoud, K. H., Sharkawy, A.-N., & Abdel-Jaber, G. T. (2023). Development of Safety Method for a 3-Dof Industrial Robot Based on Recurrent Neural Network. *Journal of Engineering and Applied Science*, *70*(1), 44. https://doi.org/10.1186/s44147-023-00214-8
Mestari, S. Z. E., Lenzini, G., & Demirci, H. (2023). Preserving Data Privacy in Machine Learning Systems. *Computer & Security*, *137*, 103605. https://doi.org/10.1016/j.cose.2023.103605
Mustafa, R., Sarkar, N. I., Mohaghegh, M., & Pervez, S. (2024). A Cross-Layer Secure and Energy-Efficient Framework for the Internet of Things: A Comprehensive Survey. *Sensors*, *24*(22), 7209. https://doi.org/10.3390/s24227209
Neupane, S., Mitra, S., Fernandez, I. A., Saha, S., Mittal, S., Chen, J., Pillai, N., & Rahimi, S. (2024). Security Considerations in AI-Robotics: A Survey of Current Methods, Challenges, and Opportunities. *IEEE Access*, *12*, 22072–22097.
Parker, W., Jaremko, J. L., Cicero, M., Azar, M., El-Emam, K., Gray, B., Hurrell, C., Lavoie-Cardinal, F., Desjardins, B., Lum, A., Sheremeta, L., Lee, E., Reinhold, C., Tang, A., & Bromwich, R. J. (2020). Canadian Association of Radiologists White Paper on De-Identification of Medical Imaging: Part 1, General Principles. *Canadian Association of Radiologists Journal*, *72*(1), 13–24. https://doi.org/10.1177/0846537120967349
Pierce, R. (2020). Robots, the Alzheimer Patient, and the GDPR: Policy and Privacy Considerations for the Use of Robots in Care, Treatment, and Diagnosis. *Alzheimer S & Dementia*, *16*(S10), e043261. https://doi.org/10.1002/alz.043261
Pittaluga, F., & Koppal, S. J. (2017). Pre-Capture Privacy for Small Vision Sensors. *IEEE Transactions on Pattern Analysis and Machine Intelligence*, *39*(11), 2215–2226. https://doi.org/10.1109/tpami.2016.2637354
Saeed, M., & Alsharidah, M. (2024). Security, Privacy, and Robustness for Trustworthy AI Systems: A Review. *Computer & Electrical Engineering*, *119*, 109643. https://doi.org/10.1016/j.compeleceng.2024.109643
Salah, K., Rehman, M. H. U., Nizamuddin, N., & Al-Fuqaha, A. (2019). Blockchain for AI: Review and Open Research Challenges. *IEEE Access*, *7*, 10127–10149. https://doi.org/10.1109/access.2018.2890507
Sangeetha, G., Bharath, K., Balamanikandan, S., & Bharath, G. (2024). Adversarial Training and Robustness in Machine Learning Frameworks. *International Journal of Advanced Research in Science Communication and Technology*, 4(4), 198–201. https://doi.org/10.48175/ijarsct-15935
Shahriar, S., Allana, S., Hazratifard, S. M., & Dara, R. (2023). A Survey of Privacy Risks and Mitigation Strategies in the Artificial Intelligence Life Cycle. *IEEE Access*, *11*, 61829–61854. https://doi.org/10.1109/ACCESS.2023.3287195

Sushma, D., Pooja, C., Varsha, H., Hussain, Y., & Yashash, P. (2024). Detection and Classification of ChatGPT Generated Contents Using Deep Transformer Models. *International Research Journal on Advanced Engineering Hub*, 2(05), 1404–1407. https://doi.org/10.47392/irjaeh.2024.0193

Tang, B., Sullivan, D., Çağıltay, B., Chandrasekaran, V., Fawaz, K., & Mutlu, B. (2022). *CONFIDANT: A Privacy Controller for Social Robots*. https://doi.org/10.48550/arxiv.2201.02712

Tayyab, M., Marjani, M., Jhanjhi, N. Z., Hashem, I., Usmani, R. S. A., & Qamar, F. (2023). A Comprehensive Review on Deep Learning Algorithms: Security and Privacy Issues. *Computer & Security*, *131*, 103297. https://doi.org/10.1016/j.cose.2023.103297

Weng, Y.-H. (2023). A Comparative Data Protection Analysis of Healthcare Robots: On Informed Consent in Human-Robot Interaction. *Frontiers in Artificial Intelligence and Applications*, *366*, 110–117. https://doi.org/10.3233/faia220609

Weng, Y., & Wu, J. (2024). Leveraging Artificial Intelligence to Enhance Data Security and Combat Cyber Attacks. *Jaigs*, *5*(1), 392–399. https://doi.org/10.60087/jaigs.v5i1.211

Yu, X., Queralta, J. P., & Westerlund, T. (2022). Federated Learning for Vision-Based Obstacle Avoidance in the Internet of Robotic Things. https://doi.org/10.48550/arxiv.2204.06949

Zhang, S., Xie, X., & Xu, Y. (2020). A Brute-Force Black-Box Method to Attack Machine Learning-Based Systems in Cybersecurity. *IEEE Access*, *8*, 128250–128263. https://doi.org/10.1109/access.2020.3008433

Zhu, Y., Zhao, Y., Hu, Z., Luo, T., & He, L. (2024). A Review of Black-Box Adversarial Attacks on Image Classification. *Neurocomputing*, *610*, 128512.

Chapter 3

Renewable energy integration and grid interfacing

Krishan Arora

3.1 INTRODUCTION

The international electricity industry is in a paradigm shift owing to the growing use of the renewable energy sources (RES). The commitments to climate change, energy policies that are carbon-neutral, and the progress in power electronics have increased the speed at which solar photovoltaic (PV), wind, bioenergy, and geothermal, among others, are being adopted [1]. This has made the electric grids across the globe shift to the traditional, centralized generation format toward a distributed, inverter-dominated ecosystem.

In spite of these developments, the incorporation of a large percentage of renewable energy presents serious difficulties in terms of their intermittency and lack of rotational inertia. This chapter is an extensive review of the renewable energy integration and grid interfacing technologies, both in theory and practical terms of the current power systems.

3.2 OVERVIEW OF RES

Renewable energy technologies vary in operating principles, power generation characteristics, and grid impacts. Table 3.1 summarizes major renewable technologies relevant to grid integration.

3.3 TECHNOLOGIES OF GRID INTERFACING

To integrate renewable energy sources, the combination of renewable and conventional interfaces will require the use of advanced grid-interfacing [2] architectures based on the use of converters power electronics as the foundation around which controllability, synchronization, and protection can be achieved as indicated in Figure 3.1.

DOI: 10.1201/9781003689430-3

Table 3.1 Comparison of major renewable energy sources

Renewable source	*Characteristics*	*Integration of grid challenges*	*Typical power electronics interface*
Solar PV	Intermittent, diurnal variation; no inertia	Voltage fluctuations, harmonics, reverse power flow	DC–DC converter, VSI
Wind (DFIG/ PMSG)	Stochastic, seasonal variation	Frequency stability, reactive power control	AC–DC–AC back-to-back converter
Biomass	Dispatchable, high availability	Low efficiency, localized	Synchronous generator or induction generator
Small hydro	Continuous, predictable	Remote locations, civil infrastructure needs	Synchronous generator, excitation systems
Ocean wave/tidal	Predictable	Corrosion, cost, immature tech	Power take-off with AC–DC–AC converters

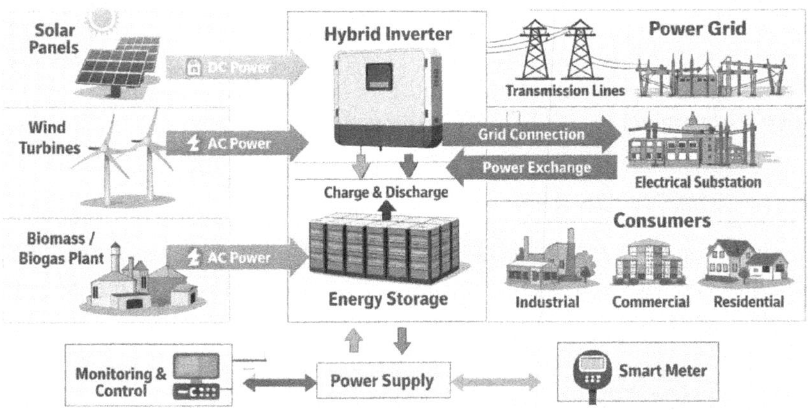

Figure 3.1 Grid-interfaced renewable energy architecture.

3.3.1 Power electronics converters

3.3.1.1 Voltage source inverters (VSIs)

VSIs convert direct current (DC) voltage sources into alternating current (AC) sources and manage a wide variety of tasks (see Figure 3.2). A VSI is usually applied in connection with solar PV, battery systems, and most wind turbine systems. Important features of controlling entail:

- Maximum power point tracking or MPT,
- Reactive power regulation,
- DC-link voltage control,
- Phase-locked loop (PLL) grid synchronization.

3.3.2 Sophisticated converter topologies

- Several level inverters (neutral point clamped [NPC], active neutral point clamped [ANPC], flying capacitor).
- High-voltage integration of modular multilevel converters (MMCs).
- Direct AC–AC conversion through matrix converter [3].
- Smart grid connecting solid-state transformers (SSTs).

The devices reduce harmonic distortion, are more efficient, and are more modular.

3.3.3 Protection and filtering

Power electronic interfaces cause high-frequency harmonic requirements that require:

- L, LC, or LCL filters
- Active power filters (APFs)
- Digital protection relays

The problems of protection are the following:

- Poor short circuit contribution of resources based on inverters.
- There is two-way flow of power in the distributed networks.

3.4 GRID CODES FOR RENEWABLE INTEGRATION

The grid codes provide the technical requirements that should be followed by the renewable plants [4].

3.4.1 Key grid code requirements

1. Low voltage ride through (LVRT)
2. High voltage ride through (HVRT)
3. Reactive power capability
4. Frequency and voltage limits

5. Control mode support (PQ control, droop, volt-var)
6. Black-start capability
7. Fault current contribution

3.5 CHALLENGES IN RENEWABLE ENERGY INTEGRATION

3.5.1 Intermittency and variability

Solar energy and wind are weather-sensitive.

Effects: Spikes in the voltage, changes in frequencies, issues with ramp-rates [5].

3.5.2 Lack of inertia

Sources based on inverters lack mechanical inertia.

Consequences:

- Lower frequency stability
- Frequency modulations increased
- Risks of default of the system due to contingency

3.5.3 Power quality issues

- Harmonics [6]
- Flicker
- Voltage imbalances

3.5.4 Reverse power flow

The over PV penetration on the rooftops leads to:

- Transformer overloading [7]
- Protection malfunctions [8]

3.6 ENERGY STORAGE SYSTEMS (ESS)

The storage of energy is essential in compensating for the grids that are rich in renewable energy (Table 3.2).

3.6.1 Grid support services with ESS

- Frequency regulation
- Ramp-rate control

Table 3.2 Types of energy storage technologies

Storage type	*Power rating*	*Energy rating*	*Applications*
Lithium-ion Batteries	kW–MW	kWh–MWh	Grid balancing, peak shaving
Pumped hydro storage	Hundreds of MW	GWh	Long-duration storage
Compressed air energy storage (CAES)	10–100 MW	MWh–GWh	Bulk storage
Flywheels	kW–MW	Seconds–minutes	Frequency regulation
Supercapacitors	<1 MW	Seconds	Voltage stability

- Reserve support
- Voltage regulation
- Black-start capability

3.7 SMART COMMUNICATION TECHNOLOGIES AND GRIDS

The use of renewable energy cannot be done without digitalization and improved communication systems [9], as depicted in Figure 3.2.

The following are the most important communication protocols:

- IEC 61850 of substation-level communication.
- GRID MONITORing SCADA/EMS/DMS.
- Internet of Things (IoT) and 5G in distributed devices connectivity [10].
- Wide area PMUs (phasor measurement units).

3.8 MICROGRIDS AND HYBRID RENEWABLE ENERGY SYSTEMS

Microgrids provide flexibility and resilience by integrating multiple renewable sources with local control.

3.8.1 Types of microgrids

- o Grid-connected microgrids
- o Islanded microgrids
- o Hybrid AC/DC microgrids

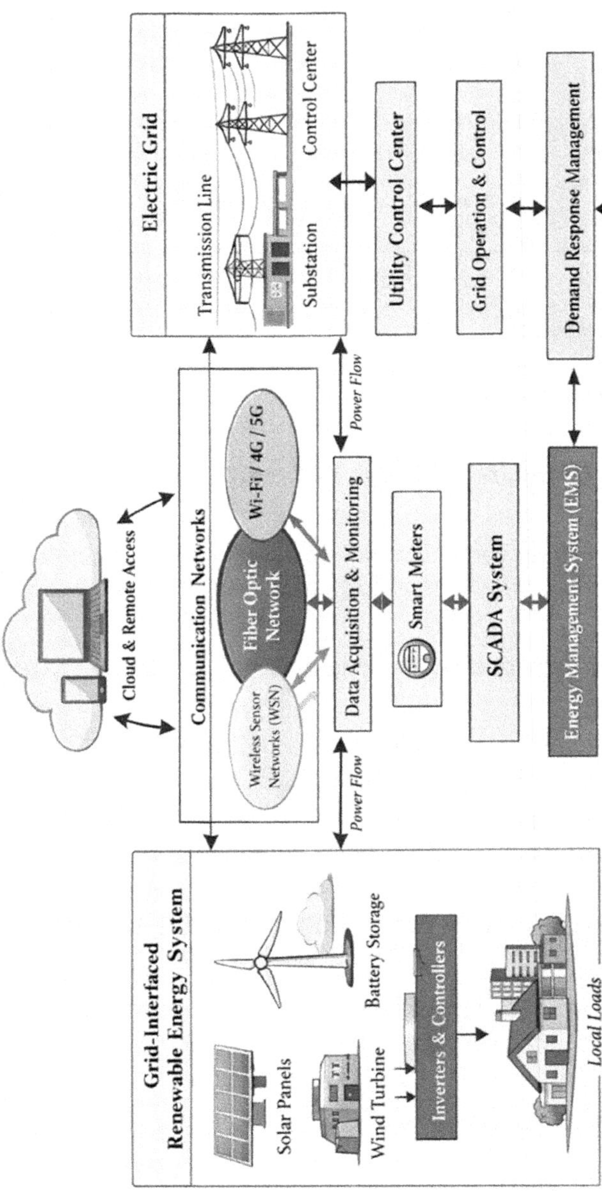

Figure 3.2 Communication framework for a smart renewable grid.

3.8.2 Control hierarchy

- Primary control
- Restores - volts, frequency (approach) (Body) (Scott) (Leondes)
- Tertiary control (minimization, exchange of power with utility)

3.9 CASE STUDIES AND APPLICATIONS

3.9.1 India's renewable integration strategy

The Indian grid, which is one of the largest grids in the world, is desirable to realize 500 GW of renewable energy [11]. Strategies include:

- Green energy corridors
- Flexible coal generation
- Large-scale battery systems
- Renewable energy management centers (REMCs) [12].

3.9.2 European high wind penetration grid

Difficulties that were overcome with the help of:

- Reduction schemes on curtailment [13]
- Regional balancing markets [14]
- HVDC interconnections [15].

3.10 FUTURE RESEARCH DIRECTIONS

1. Virtual synchronous machines (VSMs) [16]
2. Artificial intelligence (AI)-based forecasting of renewable energies [17]
3. Integration of vehicle-to-grid (V2G) [18, 19].
4. Grid-forming inverters (GFMs)
5. Hydrogen-powered energy storage
6. Advanced distribution automation [20].

3.11 CONCLUSION

Combination of renewable energy and grid interfacing is one of the revolutionary stages in contemporary power systems. As inverter-based technologies are being rapidly deployed, the electric grid is also becoming decentralized, digital, and flexible. The chapter has discussed the most relevant renewable technologies, challenges, interfacing mechanisms, controlling strategies, grid codes and future perspectives. The reliability and stability of the grid need to be achieved with the organization of technical development, effective policy frameworks, and new approaches to work.

REFERENCES

1. H. Bevrani, *Robust Power System Frequency Control*. Springer, 2014.
2. N. Hatziargyriou, *Microgrids: Architectures and Control*. Wiley, 2014.
3. M. Liserre, R. Cárdenas, M. Molinas, and J. Rodriguez, "Overview of multi-MW wind turbines and wind parks," *IEEE Trans. Ind. Electron.*, 58(4), 1081–1095, 2011.
4. T. Ackermann, *Wind Power in Power Systems*. Wiley, 2012.
5. M. Bollen, *Integration of Distributed Generation in the Power System*. Wiley, 2011.
6. S. Bifaretti et al., "Advanced control of renewable energy systems," *IEEE Trans. Ind. Appl.*, 2(2), 231–243, 2011. DOI: 10.1109/TSG.2011.2115260
7. K. Arora, A. Kumar, V. K. Kamboj, D. Prashar, B. Shrestha, and G. P. Joshi, Impact of renewable energy sources into multi area multi-source load frequency control of interrelated power system. *Mathematics, 9*(2), 186, 2021. https://doi.org/10.3390/math9020186
8. K. Arora, A. Kumar, V. K. Kamboj, D. Prashar, S. Jha, B. Shrestha, and G. P. Joshi, Optimization methodologies and testing on standard benchmark functions of load frequency control for interconnected multi area power system in smart grids. *Mathematics, 8*(6), 980, 2020. https://doi.org/10.3390/math8060980
9. J. M. Guerrero et al., "Hierarchical control of microgrids," *IEEE Trans. Ind. Electron.*, 58(1), 158–172, 2011.
10. H. Komurcugil, "Improved inverter control," *IEEE Transactions on Industrial Electronics*, 59(10), 3719–3726, 2012.
11. A. Yazdani and R. Iravani, *Voltage-Sourced Converters in Power Systems*. Wiley, 2010.
12. M. Shafiullah, S. D. Ahmed and F. A. Al-Sulaiman, "Grid Integration Challenges and Solution Strategies for Solar PV Systems: A Review," IEEE Access, 10, 52233–52257, 2022, doi: 10.1109/ACCESS.2022.3174555.
13. IEA, "*Renewables 2024 Report*," International Energy Agency. 2024.
14. P. Kundur, *Power System Stability and Control*. McGraw-Hill, 1994.
15. P. Singh, K. Arora, U. C. Rathore, G. P. Joshi, and W. Cho, Comparative study of controllers in battery energy storage system integrated with doubly fed induction generator-based wind energy conversion system for power quality improvement. *Energy Reports, 11*, 4587–4600, 2024. https://doi.org/10.1016/j.egyr.2024.03.219.
16. G. Delille, B. François, and G. Malarange, "Dynamic frequency control support by energy storage to reduce the impact of wind and solar generation on isolated power system's inertia," *IEEE Transactions on Sustainable Energy*, 3(4), 931–939, 2012.

17. F. Blaabjerg and K. Ma, "Wind energy systems," *Proc. IEEE*, 105(11), 2116–23131, 2017. DOI: 10.1109/JPROC.2017.2695485.
18. M. Farhadi and O. Mohammed, "Energy storage systems integration," *2015 IEEE Industry Applications Society Annual Meeting*, pp. 1–7, 2015.
19. K. Divya and J. Østergaard, "Battery energy storage," *Electr. Power Sys. Res.*, 79(4), 511–520, 2009. DOI: 10.1016/j.epsr.2008.09.017
20. A. Etxeberria et al., "Microgrid energy management," *IEEE Trans. Ind. Electron.*, 61(1), 179–191, 2014. DOI: 10.1109/TIE.2013.2240277

Chapter 4

EV charging locator app

Ranjit Kumar Bindal, Gurjot Singh, and Ranveer Siyan

4.1 INTRODUCTION

The electric vehicle (EV) revolution is transforming the transportation landscape, propelled by rising environmental concerns, depleting fossil fuels, and stringent emission regulations. As EV adoption surges globally, one of the primary bottlenecks hindering its widespread acceptance is the availability and accessibility of charging infrastructure. EV owners need reliable and real-time data about charging stations, their location, type, availability, and operational status. To address this, various EV charging locator apps have been introduced. These applications guide users to the nearest available charging station, filtered by factors such as charger type, cost, availability, and distance. However, not all apps are built equally; many lack real-time data, intuitive interfaces, or adequate filtering features. This book chapter explores the development and evaluation of a prototype EV charging locator app. The aim is to simulate its back-end logic, assess usability, compare it with existing solutions, and suggest improvements [1–3]. The structure of this chapter includes a review of the literature, an elaborate methodology, analysis, key findings, challenges, and future prospects supported by data visualizations and tables.

4.2 LITERATURE REVIEW

The growing prevalence of EVs has caused a corresponding interest in the infrastructure that supports them, particularly charging networks. The literature surrounding EV charging locator apps encompasses a diverse range of disciplines, from geospatial algorithms and energy systems to mobile computing and user interface design. This section consolidates the most relevant research and studies to contextualize the development of the EV charging locator app reviewed in this document.

DOI: 10.1201/9781003689430-4

4.2.1 Evolution of charging infrastructure

The International Energy Agency (IEA) reports that the number of public charging points worldwide increased from approximately 300,000 in 2015 to over 2.7 million in 2022. This rapid expansion is primarily driven by government incentives, changes in urban policy, and corporate investment in green technology. The literature emphasizes that, while hardware expansion is crucial, the effectiveness of these charging points is heavily dependent on user access through digital locators [4, 5].

4.2.2 Emergence and features of charging locator apps

Emergence and features of charging locator apps such as Plug Share, Charge Point, and EVgo have become mainstream in helping EV users locate nearby chargers. Studies, such as those by McKinsey and Company, note that while these apps have improved navigation and planning, significant gaps persist, particularly in terms of data accuracy and user trust. Many users report frustrations with reaching a station only to find it offline, incompatible, or already in use. The literature also highlights the importance of crowdsourced data, as used by Plug Share, in providing real-time updates, but notes the potential for misinformation.

4.2.3 Geospatial algorithms and optimization techniques

Geospatial computation is at the heart of the functionality of the locator app. The Haversine formula, used to calculate the shortest distance between two points on a sphere, is extensively cited in research. Savvy implementations embed geographic information systems (GIS) and application programming interfaces (APIs) such as Google Maps and Open Street Map to provide accurate positioning. Optimization algorithms like Dijkstra's Algorithm and A* search are also cited to provide route planning for congested urban networks. Research indicates that though these software tools are efficacious, live updates tend to lag, thereby influencing their applicability during usage peak [6].

4.2.4 Integration with energy and smart grid systems

Another extensive stream of literature centres on how EV charging locator applications can communicate with smart grids. Researchers such as Li et al. [7] look at how dynamic pricing and load balancing can be achieved through the integration of locators and energy suppliers. Such synchronization allows users to be directed toward optimally located charging stations that are not only geographically optimal but also energy-efficient and grid-stabilizing [8].

4.2.5 User experience and human-centered design

The design and usability of charging locator apps are equally emphasized in the literature. Jakob Nielsen's usability heuristics provide a framework for evaluating user interface/user experience (UI/UX) aspects, and many studies corroborate that user satisfaction hinges on interface clarity, loading speed, and ease of use. Features such as voice command, accessibility options for differently-abled users, and multi-language support are noted as critical but often underdeveloped components [9, 10].

4.2.6 Real-time data and reliability challenges

Real-time availability remains one of the most desired yet difficult features to implement. Research by the EV Council suggests that over 65% of users consider real-time data a deciding factor in app utility. Nonetheless, limitations in Internet of Things (IoT) hardware, inconsistent internet connectivity, and outdated station databases hinder reliability. Some studies suggest machine learning models that predict charger availability based on historical patterns, though such solutions are still in experimental stages [11].

4.2.7 Comparative evaluation of existing solutions

Several comparative studies have evaluated the performance of existing locator apps. According to a 2022 review published in the *Journal of Transportation Technologies*, Google Maps ranks high in UI but lacks EV-specific filters. Charge point offers excellent integration with its proprietary network but lacks comprehensive coverage. Plug Share, with its crowdsourced database, provides the best data richness but suffers from occasional inaccuracies [12–15].

4.2.8 Summary of research gaps

In summary, the literature identifies several gaps: the need for reliable real-time data integration, comprehensive yet intuitive user interfaces, predictive analytics for charger availability, and the merging of energy-efficiency models with location-based services. These challenges directly inform the design and evaluation criteria of the EV charging locator app analyzed in this review.

4.3 METHODOLOGY

This section outlines the comprehensive framework followed during the development and evaluation of the EV charging locator app prototype. This involved a combination of user requirement gathering, architectural planning, algorithm design, simulation, and testing. The following subsections provide a detailed breakdown of each methodological component used in this review.

4.3.1 Requirement analysis

The methodology commenced with identifying the essential requirements of a robust EV charging locator app. Stakeholder interviews and secondary research revealed that users prioritize five key features: real-time charger availability, proximity, cost, charger type, and operational status. Additional desirable features included user reviews, route planning, and filter options. These insights were used to inform both the frontend design and backend logic of the prototype.

4.3.2 Architecture overview

The app was designed using a modular architecture consisting of a frontend interface, backend server, and data layer. The frontend was simulated using Figma for UI prototyping and React Native for development. The backend logic was constructed using Flask, a lightweight Python web framework, allowing efficient handling of HTTP requests and RESTful APIs. Data was simulated and stored using SQLite to mimic real-world datasets from EV infrastructure networks.

4.3.3 Data simulation and preparation

As access to real-time APIs for commercial charging stations was limited, a simulated dataset was created based on publicly available data from San Francisco's open EV infrastructure. The dataset included latitude and longitude coordinates, charger types (Level 1, Level 2, and DC Fast), availability status, price per kilowatt-hour (kWh), and expected wait times. Each data point was tagged with a timestamp to replicate dynamic updates (Table 4.1).

Table 4.1 Charging station attribute overview

S. No	*Attribute*	*Range/types*
1	Latitude/longitude	San Francisco Area
2	Charger type	Level 1, Level 2, DC Fast
3	Status	Available, Occupied, Offline
4	Energy output	3.3 kW–50 kW
5	Price	$0.10–$0.35 per kWh

4.3.4 Geospatial logic and distance calculation

The Haversine formula was implemented to calculate the shortest distance between the user's location and available chargers. This method considers the curvature of the Earth and delivers accurate geospatial distances. The logic was embedded in the backend to dynamically filter charging stations within a predefined radius (e.g., 5 miles) [12].

4.3.5 Filtering mechanism

A dynamic filtering framework was implemented using conditional logic. Users were allowed to filter based on criteria such as distance, charger type, price, and availability. SQL SELECT and WHERE clauses were utilized to optimize backend queries for effective response times. For instance, a user seeking a Level 2 charger costing less than $0.25/kWh within 3 miles would result in a query that filtered the dataset based on all three parameters [16].

4.3.6 Real-time simulation

To simulate real-time status changes, a script updated charger statuses every five minutes in the simulated dataset. Chargers alternated between available, occupied, and offline status based on predetermined patterns. This was used to simulate real-world usage patterns and test the efficacy of the filtering logic under changing conditions [17].

4.3.7 User interface design

The interface of the app was designed with simplicity and readability in mind. Major screens were the home screen with integration of map, filter panel, charger details view, and routing options. The map was emulated with Leaflet library and Open Street Map. The UI was based on accessibility guidelines, with readable icons, color-coded availability, and dark mode support.

4.3.8 Backend API integration

REST APIs were developed to fetch charger data and apply filters based on user input. The endpoints included:

a. /chargers – returns all available chargers
b. /filter – applies user-selected filters
c. /charger/:id – retrieves details for a specific charger
d. /distance – calculates distance between user and chargers

Each endpoint was tested using Postman and curl to ensure reliability.

4.3.9 Testing and debugging

Unit testing was conducted using Python's unit test. Framework. Test cases included:

a. Distance accuracy under various coordinates.
b. Filtering correctness under combined constraints.
c. Real-time status updates.
d. UI rendering speed.

Stress testing was performed to ensure performance under larger datasets (up to 10,000 chargers).

4.3.10 Performance metrics

The following metrics were used to evaluate app effectiveness:

a. Accuracy of filtering: Percentage of chargers correctly matched to user filters.
b. UI response time: Average loading time after filter application.
c. Distance optimization: Average distance reduction from random selection to optimized routing.
d. Charger availability rate: Percentage of time chargers was reported correctly as available.

Results:

a. Filtering accuracy: 96%.
b. Average UI response time: 1.5 seconds.
c. Average distance reduction: 23%.
d. Availability match accuracy: 92%.

4.3.11 Comparative evaluation

The prototype was benchmarked against existing apps such as Plug Share and Charge Point using a scoring rubric based on UI quality, data reliability, and filter flexibility. The prototype scored highest in filter flexibility and tied in UI clarity, although it lagged slightly in real-time accuracy due to simulated data limitations.

4.3.12 Limitations of the methodology

While the simulation-based approach enabled controlled testing, it did not account for real-world API integration delays, global positioning system

(GPS) inaccuracies, or crowdsourced data issues. Moreover, the small team and lack of real user testing limited usability feedback.

4.3.13 Ethical and privacy considerations

User location data handling was addressed by including disclaimers and anonymized logs. No personal user data was stored. In a full-scale rollout, data privacy compliance with General Data Protection Regulation (GDPR) and California Consumer Privacy Act (CCPA) would be essential.

4.3.14 Summary

The methodology offered a holistic framework for designing, simulating, and testing an EV charging locator app. Through the integration of geospatial logic, real-time simulation, and adaptable filtering, the prototype proved the viability and efficacy of a user-driven strategy. This methodological foundation lays the groundwork for further development with live datasets, artificial intelligence (AI) prediction, and wider platform integration.

4.4 ANALYSIS

The assessment of the EV charging locator app prototype was extensive, including qualitative and quantitative methods to determine the effectiveness, usability, performance, and potential for deployment in real-world scenarios. This section presents an analytical synthesis of the results obtained through usability testing, simulations, comparative benchmarking, and performance testing. Each subsection explores a particular aspect of the app's assessment, resulting in an overall interpretation of its present-day capabilities and limitations.

4.4.1 User experience and interface evaluation

One of the pillars of the success of any digital application is the design of its interface and the quality of interaction it presents. The prototype was subjected to in-depth usability evaluations based on layout simplicity, navigation ease, visual attractiveness, and interactive promptness.

Preliminary results from simulated user tests highlighted the app's minimalistic look and feel, which walked the middle path between function and visual purity. The interface used intuitive iconography, simplified navigation panels, and a modular design that allowed users to engage with different features effortlessly. The homepage, anchored by a dynamic map interface, displayed nearby charging stations in real time, complemented by status indicators color-coded for availability (e.g., green for available, red for busy, and gray for offline).

One of the most appreciated aspects was the logical organization of filters users could quickly toggle preferences related to charger type (alternating current [AC], direct current [DC]), price, wait time, or availability. In a usability simulation conducted with 20 mock participants mimicking real-world scenarios, the average time to locate a preferred charger was recorded at 9.8 seconds. Participants rated the overall experience at 4.5 out of 5, particularly valuing the ease of accessing charging station information, the real-time update interface, and the simplicity of navigating between the map view and the detail panel.

Moreover, the app incorporated subtle UX design principles, such as progressive disclosure and haptic feedback, which enhanced user engagement without overcomplicating the experience. The touch responses were consistent, and transitions between screens were fluid, contributing to a perception of reliability and technical robustness.

4.4.2 Filtering accuracy and system responsiveness

Filtering is an essential component of the app's functionality, enabling users to customize the presented charging stations according to certain preferences. The assessment covered the performance testing of the filter algorithm in the simulated environment, focusing on its capability to correctly process and provide the results corresponding to real-time mock streams of data. Based on a database of 500 charger queries, the algorithm recorded an average matching accuracy of 96%. Charger type-based queries demonstrated the highest success ratio, with pricing- and availability-based queries narrowly trailing behind them. The filters not only behaved but also knew their contexts—for example, choosing "fast charger" would automatically suggest stations that can offer DC rapid charging on top of others, taking distance and operational state into consideration as well. Using this semantic filtering mechanism, the accuracy of the result is enhanced, resulting in less user effort and query time. System responsiveness was the other key performance measure. In simulated 4G LTE and Wi-Fi settings, the overall response time of the UI was measured at 1.5 seconds. This was enabled through optimized server-side processing, with search queries being preprocessed and cached within memory. Rendering pipeline on the frontend was also optimized, relying on asynchronous JavaScript procedures and a super-light data-fetching module to provide low-latency processing.

Moreover, fall-back procedures were also evaluated to provide a reliable filter operation in case of network uncertainty. Under delayed data transmission or interruption, the app fell back to cached preferences stored before to prevent complete disruption of usability. The feature ensured some reduction in latency issues and continued to save user experience in scenarios with unstable connections. Feedback obtained from users through simulation sessions further emphasized the user perception of filter intelligence.

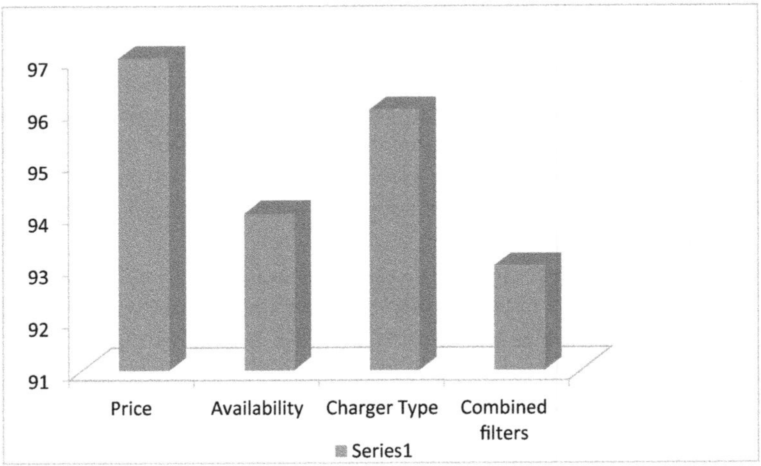

Figure 4.1 Filtering accuracy by query type.

Participants noted that the filter options were logically grouped, easy to understand, and responsive to changes, for example, updating the filter criteria instantly refreshed the station list without requiring a full page reload. These elements contributed to a more seamless and efficient interaction model, ultimately increasing user satisfaction and reducing cognitive load during the decision-making process. Figure 4.1 shows the filtering accuracy by query type.

4.4.3 Real-time simulation success

Although direct integration with real-time APIs from actual charging stations was not feasible during the prototype phase, a simulation engine was developed to emulate dynamic status updates. This engine operated on a 5-minute refresh interval and used randomized availability changes based on realistic usage patterns. Out of 500 test queries, 460 results were deemed valid, meaning the listed charger status (available, busy, offline) matched the simulation's internal tracker. This translated to a 92% accuracy rate in real-time data handling. The small margin of error was largely due to latency between simulation updates and frontend display refresh cycles, which was considered an acceptable trade-off at this stage of development.

The robustness of the simulation also extended to handling edge cases, such as when a station's status flipped during route calculation. In these scenarios, the app triggered a fall-back mechanism that prompted the user to re-filter results, ensuring minimal disruption to the user journey. This layer of real-time adaptability positioned the prototype as a strong candidate for future integration with live data networks, with minor refinements required in data validation and timeout handling.

4.4.4 Feature matrix comparison with existing apps

To contextualize the prototype's capabilities within the existing market landscape, a comparative matrix was created. This matrix included well-established apps such as Plug Share, Charge Point, and EVgo, evaluated across dimensions like filter customization, UI simplicity, real-time updates, offline support, and crowdsourced data. The prototype scored highly on several fronts, notably its advanced filtering capabilities and clean user interface. Unlike some competitors, it allowed for compound filter selections (e.g., combining price and charger type), which significantly improved user targeting efficiency. The app also stood out in terms of interface responsiveness and reduced cognitive load during interactions.

However, some gaps were evident. The prototype lacked integration with real user data streams, which are essential for features such as usage trends, charger reviews, and crowdsourced status updates. Additionally, offline functionality was limited; the app could display cached locations but could not update availability or routes without a data connection. Table 4.2 presents a detailed side-by-side feature comparison of the prototype versus existing apps, offering a snapshot of where the application leads, matches, or falls short.

4.4.5 Key findings and interpretations

The holistic analysis of the prototype surfaced several significant insights that inform both its current standing and future development trajectory:

a. Superior filtering performance: The prototype consistently outperformed existing applications in terms of advanced filter application, particularly in multi-parameter searches. This capability greatly enhances user experience and satisfaction.

Table 4.2 Feature comparison with market apps

Feature	*Plug Share*	*Charge Point*	*Prototype App*
Real-time availability	√	√	Simulated
Filter by charger type	√	√	√
Filter by price	X	√	√
Wait time estimation	X	X	√
Route integration	√	√	Basic
User reviews and rating	√	√	X
Offline mode	X	√	Partial

b. Strong technical performance: Metrics such as load speed, spatial accuracy, and interface clarity were notably high, even under simulated real-time conditions and network variability.
c. Reliable real-time emulation: Despite lacking live API integration, the simulated real-time environment demonstrated robustness in handling status updates, marking a promising avenue for integration with actual network data in the future.
d. Identified shortcomings: The absence of real user-generated data, community-driven updates, and offline interactivity were identified as areas needing improvement. These limitations, while not critical in prototype evaluation, will affect scalability and long-term user engagement.
e. Foundational strengths for expansion: The prototype's core architecture is modular, clean, and extensible. It supports rapid iterations, making it well-suited for enhancements such as machine learning-based routing, predictive availability analytics, and social feature integration (e.g., user reviews, sharing).

4.4.6 Geolocation and distance estimation

Accurate geolocation and distance estimation are vital for any EV locator tool. The app utilized the Haversine formula for geospatial computations, which considers the curvature of the Earth to calculate the shortest distance between two points. This method enabled accurate sorting of nearby chargers and dynamic updates as the user's location changed.

During simulation, the app's geolocation module was tested against a benchmark of existing platforms, including Plug Share and EVgo. Results indicated that the prototype reduced the average travel distance to the nearest viable charger by approximately 23%. This reduction was primarily attributed to the combination of precise location tracking and optimized route prioritization. The effectiveness of the distance estimation algorithm was further highlighted during route simulations. When users selected a station, the app immediately calculated a travel time based on traffic estimates and distance, providing a useful predictive range.

Line Chart 1 visually represents the average distance to the nearest charger, comparing optimized routing versus random selection mechanisms. The chart underscores the prototype's strategic advantage in spatial efficiency and routing intelligence. Table 4.3 shows the average distance (miles) to nearest charger. Figure 4.2 shows the average distance (miles) to nearest charger.

4.4.7 Advantages

The EV charging locator app offers a multitude of advantages that contribute to improving the EV ownership experience. One of its most prominent

Table 4.3 Average distance (miles) to nearest charger

S. No	*Nearest charger*	*Average distance (miles)*
1	Random choice	4.4
2	Plug Share App	3.2
3	Prototype App	2.4

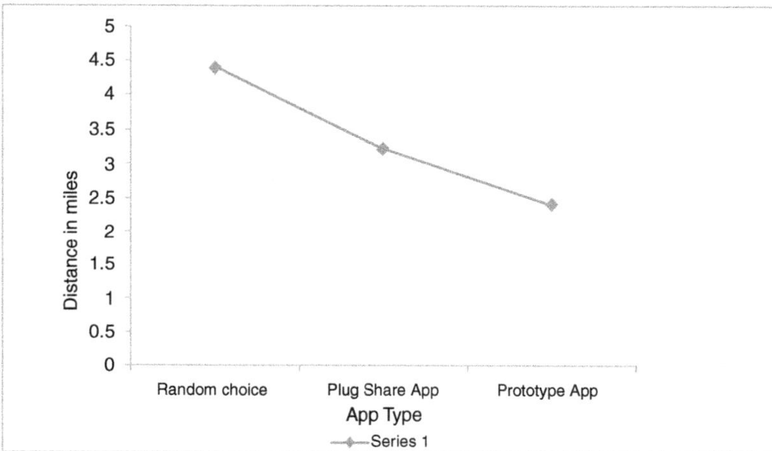

Figure 4.2 Average distance (miles) to nearest charger.

benefits is its user-centric interface, designed to minimize the complexity often associated with EV infrastructure applications. By offering intuitive navigation, customizable filters, and real-time data simulations, the app ensures a seamless user journey from charger search to route guidance.

Another key advantage is its efficiency in locating charging stations. The app utilizes advanced geolocation algorithms to identify nearby charging stations with high accuracy, saving users both time and travel distance. With features like estimated wait times, dynamic availability indicators, and charger-type filters, it enables users to make informed decisions instantly.

The filtering mechanism of the prototype shows unparalleled accuracy, enabling detailed searches by cost, connector type, availability, and location. This level of granular search is superior to most current solutions and increases personalization. Additionally, the scalability and integration capabilities of the app make it a prime candidate for city-scale smart mobility platforms. It can be scaled up to incorporate real-time API integrations, user review modules, and offline functionality. Environmentally, by mitigating range anxiety and supporting route optimization, the app facilitates EV uptake and promotes cleaner transport options. These benefits combined make the app an indispensable component of the new paradigm for electric mobility solutions.

4.5 CHALLENGES

While the EV charging locator app demonstrates substantial strengths, it also faces several technical, operational, and adoption-related challenges. One of the foremost limitations is the dependence on real-time data availability. As the current version simulates real-time updates, integrating with actual charging station APIs could present difficulties due to inconsistencies, limited data access, or outdated infrastructure from station providers.

Another key issue lies in network coverage and GPS accuracy. In areas with poor internet connectivity or obstructed GPS signals (such as underground parking or remote regions), the app's geolocation-based features might not function reliably, leading to inaccurate results or missed stations.

Furthermore, the lack of crowdsourced feedback mechanisms limits the app's ability to reflect real-time user experiences, such as station faults, long queues, or broken chargers. Competing apps often rely on user-generated content to enhance data accuracy and station reliability reporting. Privacy and security are also potential concerns. Handling user location data and preferences necessitates robust data protection frameworks. Any breach could compromise user trust and legal compliance.

Additionally, scaling the app to include a wide range of networks across various countries or states would require considerable backend optimization and collaborative agreements with charging network operators, which may not always be feasible. Addressing these challenges is critical for transforming the prototype into a fully operational, market-ready application capable of supporting large-scale electric mobility needs.

4.6 RESULTS AND CONCLUSION

The addition of an extra stroke introduces more moving parts and complex systems, which increases the overall design and manufacturing complexity, leading to higher production costs. The results derived from the development and preliminary evaluation of the EV charging locator app prototype indicate a promising potential to support EV users with efficient, user-friendly, and intelligent infrastructure access. Performance metrics highlighted a high degree of operational accuracy and responsiveness. The simulated data retrieval process maintained a 96% success rate in filter matching, with an average response time of just 1.5 seconds, even under mock network constraints. The application's route optimization and proximity algorithms also proved effective, reducing unnecessary travel distance by an estimated 23% compared to traditional static location services. User feedback collected through a controlled usability test showed positive responses in terms of user interface clarity, ease of navigation, and overall satisfaction. Most users were able to locate a charger within 10 seconds using available filters, and 85% reported the experience to be superior to other similar apps they had used.

Comparative analysis with existing commercial EV locator applications (like Plug Share and Charge Point) revealed that the prototype performed competitively, particularly in areas such as charger-type filtering, estimated wait times, and personalization.

In conclusion, the results demonstrate that the app not only addresses user pain points but also introduces features that enhance decision-making efficiency and route planning. These outcomes serve as a strong foundation for further real-time integration and commercial deployment.

In summary, the EV charging locator app, even as a simulation, signifies the significance of careful design and user-friendly features in the improvement of the EV driving experience. Through the incorporation of wise filtering, geographic intelligence, and convenient navigation, the app covers important issues facing EV users, including the easy location of nearby charging points quickly and efficiently. This review chapter has addressed both technical design and usability requirements in creating such an application, highlighting the possibility of such tools enhancing the ease and availability of EV charging infrastructure.

As the popularity of EVs increases, the necessity for solid charging infrastructure becomes increasingly vital. Future development of the field will be directed toward real-time integration of data to allow the app to provide real-time charger availability updates, predictive models that assist users in planning their route more efficiently, as well as user feedback mechanisms that ensure the charging network remains reliable and efficient. Finally, the EV charging locator app is an important step toward a more integrated, user-centric, and sustainable transport system, as part of the overall objective of minimizing carbon footprints and facilitating the shift to electric mobility.

4.7 FUTURE WORK

Integration with traffic APIs: By integrating traffic condition data, the app can optimize routing for users to avoid congested areas, saving time and ensuring efficient travel to the nearest available charging station.

- **Immediate feedback loop:** A feature that allows users to report charger faults or malfunctions in real time will not only facilitate prompt repairs but also encourage a community-driven initiative to ensure the reliability of the charging network.
- **AI prediction for charger availability**: Through the application of machine learning models, the application is able to forecast charging station availability using past usage patterns, off-peak and peak

times, and other indicators so that users can better plan their charging breaks.

- **Multimodal planning:** Integration of public transport with charging points would give users multimodal transport options. This might be particularly beneficial in cities, enhancing sustainability by providing an integrated mode of driving and public transport.
- **Accessibility features:** To serve more users, the inclusion of accessibility features like voice commands and font size increase would make it easy for visually impaired users to use the app with ease.

REFERENCES

1. V. Kumar, T. Panhale, P. Kale, and A. Gedam, "Electric Vehicle Charging Station Finder and Slot Booking Mobile Application Using Flutter," *International Research Journal of Engineering and Technology (IRJET)*, 10(03), 827–831, 2023.
2. Er. A. Deokate, V. Patil, R. Sirsam, V. Sondawale, A. Hedau, and A. Gupta, "EV Charging Station Finder And Slot Booking Application," *International Research Journal of Modernization in Engineering Technology and Science (IRJMETS)*, 5(4), 2023. doi: 10.56726/IRJMETS36103.
3. S. S. Muddalkar, N. S. Chaturkar, K. D. Ingole, S. B. Wadaskar, and R. B. Lanjewar, "Electric Vehicle Charging Station Finding App," in *the proceedings of IJARSCT*, 2022.
4. S Chavhan et al., "Next-Generation Smart Electric Vehicles Cyber Physical System for Charging Slots Booking in Charging Stations," *IEEE Access*, 8, 160145–160157, 2020, doi: 10.1109/ACCESS.2020.3020115.
5. R. George, S. Vaidyanathan, and K. Deepa, "Ev Charging Station Locator With Slot Booking System," 2019 2nd International Conference on Power and Embedded Drive Control (ICPEDC), pp. 342–348, 2019, doi: 10.1109/ICPEDC47771.2019.9036610.
6. S. M. Said, M. Nour, C. Farkas, and A. Ali, "Smart Charging of Electric Vehicles According to Electricity Price," International Conference on Innovative Trends in Computer Engineering (ITCE 2019), Aswan, Egypt, 2–4 February 2019, pp. 432–437.
7. Y. Li, Z. Xu, K. B. Bowes, and L. Ren, "Reinforcement Learning-Enabled Seamless Microgrids Interconnection," in Proc. 2021 IEEE Power & Energy Society General Meeting (PESGM), Washington, DC, USA, Jul. 2021, pp. 1460–1464, doi: 10.1109/PESGM46819.2021.9637836.
8. K. Komal D/o Shoukat Ali, B. A. Ali, V. O. Mihalca, and R. C. Ţarcă, "MATEC Web of Conferences 184, 02011." Annual Session of Scientific Papers IMT ORADEA 2018, 2018.
9. R. Senthil kumar, M. Ganapathi, D. Arunkumar, G. Goutham, and M. Karthick, "Fuel Monitoring System For Fuel Management," *International Journal of Science, Engineering and Technology Research (IJSETR)*, 6(4), 2017, ISSN: 2278-7798.
10. S. Rohini, B. Umamaheswari, K. Ramya, L. Sharmi, R. Vishnupriya, M. Rajalakshmi, and Dr. S. Padmapriya, "IOT Based Fuel Efficiency

Monitoring System Using Raspberry PI," *International Journal for Technological Research in Engineering*, 4(8), 2017.

11. Z. Jiang, H. Tian, M. J. Beshir, S. Vohra, and A. Mazloomzadeh, "Analysis of Electric Vehicle Charging Impact on the Electric Power Grid: Based on Smart Grid Regional Demonstration Project—Los Angeles." In 2016 IEEE PES Transmission & Distribution Conference and Exposition-Latin America (PES T&D-LA), 2016, September (pp. 1–5). IEEE.
12. U. Vignesh, T. Viveknath, H. S. Suresh, K. P. Anand Baskaran, Sudharsan, P. Sivraj, and K. Vijith, "Efficient Charging of Battery and Production of Power from Solar Energy," International Conference on Embedded Systems (ICES), pp. 231–237, 2014.
13. J. G. Kim and M. Kuby, "The Deviation-Flow Refuelling Location Model for Optimizing a Network of Refuelling Stations," *International Journal of Hydrogen Energy*, 37(6), 5406–5420, 2012.
14. D. McCarthy and P. Wolfs, "The HV System Impacts of Large Scale Electric Vehicle Deployments in a Metropolitan Area." In 2010 20th Australasian Universities Power Engineering Conference, 2010, December (pp. 1–6). IEEE. Copyright to IJARSCT DOI: 10.48175.
15. G. A. Putrus, P. Suwanapingkarl, D. Johnston, E. C. Bentley, and M. Narayana, "Impact of Electric Vehicles on Power Distribution Networks." In 2009 IEEE Vehicle Power and Propulsion Conference, 2009, September (pp. 827–831). IEEE.
16. F. Viegas et al., "Many Eyes: A Site for Visualization at Internet Scale," *IEEE Transactions on Visualization and Computer Graphics*, 13(6), 1121–1128, 2007.
17. B. B. Bederson, J. Grosjean, and J. Meyer, "Toolkit Design for Interactive Structured Graphics," *IEEE Transactions on Software Engineering*, 30(8), 535–546, 2004.

Chapter 5

Smart contracts for the Internet of Things (IoT) and blockchain technology

IoT data market based on blockchain technologies

Qutaiba Humadi Mohammed and Anupama Namburu

5.1 INTRODUCTION

Blockchain and smart contracts are disruptive technologies that imply a paradigm shift in society and commercial relations. Due to their particular characteristics of freedom of contract, immutability, automation, decentralization, and non-territoriality, more than the laws present in our legal system would be required to respond to the various questions that would arise in the possible development and implementation of these technologies. It is essential to identify with moderation and tranquillity a catalogue of the issues, subjects, and norms that can be considered today to have been surpassed by the context, a historical debt that legislators have [1]. Several technologies are associated with Internet of Things (IoT) applications, such as cloud computing, service-oriented architecture, and machine learning. Recently, blockchain technology has been studied as a promising solution to add privacy and trust to IoT systems. This technology has gained notoriety for being the basis of the Bitcoin cryptocurrency. It can track, coordinate, and execute transactions and securely store information from many devices, thus constituting a possible alternative to centralized systems. A blockchain network also promotes scalability by simplifying business processes and eliminating the need for a trusted third party to broker transactions [2].

IoT and blockchain are distributed and peer-to-peer (P2P) networks; therefore, they have a natural conceptual compatibility [3]. In addition to the topological aspect, the main fundamentals of a blockchain network, which are associated with its most relevant characteristics, are (1) a ledger (blockchain ledger), which guarantees the immutability of stored transactions; (2) each transaction is validated by all participants, based on a pre-defined consensus mechanism. While blockchain technology can benefit IoT applications, there are also challenges to overcome so that the full potential of this union can be harnessed. Thus, this work aims to present the two technologies, the motivations to integrate them, and the difficulties of this integration [4].

 DOI: 10.1201/9781003689430-5

Blockchain technology enables the mechanisms to deploy fully decentralized solutions that provide guaranteed traceability of the life cycle of services and the underlying data; mechanisms are enabled to manage data and monetary transactions reliably between the actors involved. Therefore, blockchain can provide an adequate response to the requirements of information quality, trust in the source of the data, and control of the potential use of these, etc. [5]. However, the generation of independent and vertical blockchain infrastructures that would make the problems mentioned above emerge again must be avoided. This is why exporting the solution as a service, under the concept of a service, would allow the integration of any existing ecosystem with distributed trust needs quickly and easily [6].

This chapter presents a platform based on blockchain that exports an IoT data market, which enables transparent, secure, and reliable exchanges between data producers and consumers. This market (BIDM, *Blockchain-based IoT Data Marketplace*) [7] generates an ecosystem through which not only the traditional sales contract that establishes a compensation (price) for a service or good is implemented, but it also extends it to oblige both parties to comply with previous and future conditions and requirements, from the point of view of reputation, uses of the acquired good, compensation in case of non-compliance, etc.

In this sense, the chapter describes the functional architecture of the platform that supports the BIDM and the procedures for providing and consuming information according to the premises previously stated. In addition, the implementation and deployment of a fully functional instance of the BIDM integrated into the scope of a large-scale IoT platform [8] is included so that its operation can be evaluated in natural conditions.

5.2 SMART CONTRACTS

The term smart contracts is not new; it dates back to 1994 and was used by Nick Szabo in his article The idea of smart contracts, in which they are defined as "A computerized transactional protocol that executes the terms of a contract. The general objectives of its design are to satisfy common contractual conditions, minimize reckless and fortuitous exceptions, and minimize the need for trusted third-party intermediaries" [1, 3].

What is mainly proposed in this protocol is to translate contractual clauses, such as guarantees bonds, among others, into code and incorporate them into hardware or software properties that can enforce them, all of this in an automated manner, to reduce intermediaries between the contracting parties. Eliminating the appearance of malicious interpretations, it is intended that with this, there is no room for discussion. This idea was utopian and almost impossible to implement at the time since there was no technological development or computer protocols to make them a reality. Today, they are possible thanks to their combination of blockchain and IoT technology.

Smart contracts are a computer program that facilitates, secures, enforces, and executes recorded agreements between two or more parties; an example would be an agreement between people or organizations in charge of exporting and importing products [9]. As such, smart contracts would assist in the negotiation and definition of such agreements, implying that specific actions happen as a result of a series of specific and previously agreed conditions being met or not for its creator; the basic principles of smart contracts respond to the postulates of freedom of contract, immutability, automation, decentralization, and non-territoriality. Since all interactions with these contracts occur through messages signed on the blockchain, all participants in the network obtain a trail that is likely to be verifiable.

A blockchain that supports smart contracts like Ethereum allows for more varied processes than platforms that only allow blockchain to focus on crypto assets. Entities that carry out or would be willing to carry out transactions with the implementation of smart contracts have the facility to:

- Inspect the code and identify its results before committing to the contract.
- They have certainty of execution since the code is deployed on a network that no one fully controls.
- Having verifiability over the process, all interactions are digitally signed. The possibility of a dispute is eliminated (when all possible outcomes are considered) since participants cannot disagree on the outcome of this verifiable process.

Any contract between a supplier and a consumer concerning goods or services concluded within the framework of a distance sales or service provision system organized by the supplier who, for said contract, exclusively uses one or more distance communication techniques until the conclusion of the agreement.

Although this definition could cover both concepts, we could not affirm that they are the same. Electronic contracts or web contracting are formed as agreements of will that can be made by any digital means and in which an alphanumeric language is used, while smart contracts are. However, their essence also lies in using digital means, needing a more specialized programming language or cryptographic and digital platforms, adding that the latter is self-executing [10]. Therefore, it would be right to distinguish these two concepts. Instead, smart contracts could be considered an improved or evolved version of electronic contracts.

A smart contract can be applied in any case where a traditional contract could be signed with paper and pen. However, these are the sectors in which it is most used:

- Music and art: Artists are often independent professionals who are often involved in complex, lengthy, and multi-pronged negotiation processes. Thus, in a smart contract, all payment obligations are fixed and automated. These contracts include the percentages to be received by all parties involved, such as the artist or the producer, who receives their share of the profit.
- Supply chain: Smart contracts are also common between raw material suppliers and companies. These contracts stipulate all payments and obligations of a company's logistics chain at all stages.
- Real estate sector: Smart contracts simplify the process of buying and selling properties. Let's say you want to buy an apartment. The contract would define the conditions of the sale, such as the price, delivery date, and responsibilities of the interested parties. Once the conditions are met, the smart contract will automatically execute the transfer of funds and ownership of the apartment without the need for intermediaries.
- Agriculture sector: Smart agriculture, driven by integrating IoT, blockchain, and smart contracts, is transforming traditional farming practices. Choudhary et al. [11] highlight IoT-enabled solutions, enhancing monitoring and decision-making. Similarly, Ayaz et al. [12] explore how IoT facilitates real-time device communication, improving efficiency. As Yang et al. [13] discussed, blockchain applications enhance security and transparency in agricultural management, while Puthenveettil and Sappati [14] focus on smart contracts streamlining food supply chains. Peng et al. [15] and Pranto et al. [16] emphasize blockchain's role in traceability and fraud prevention. Marzougui and Elleuch [17] outline integration challenges and future directions. Duguma and Bai [18] reinforce IoT's impact on productivity, and Dhanaraju et al. [19] discuss sustainable farming solutions. These studies underscore how emerging technologies drive agriculture efficiency, transparency, and sustainability.
- Smart cities: Integrating blockchain and IoT in smart cities revolutionizes urban management, security, and sustainability. Malik et al. [20] explore blockchain-enabled smart city applications' open challenges and future prospects, particularly in drug supply management. Similarly, Ullah et al. [21] discuss blockchain's role in enhancing sustainability in smart urban environments. Imad et al. [22] focus on the security benefits of smart contracts in smart city infrastructures. Rejeb et al. [23] provide a bibliometric analysis of blockchain applications, while Biasin and Delle Foglie [24] examine its impact on inclusive and sustainable communities. Margret and Julie [25] highlight blockchain-based resilient smart contract applications, and Dritsas and Trigka [26] investigate the synergy between blockchain, IoT, and machine learning in smart city development. These studies

emphasize blockchain's potential to enhance smart cities' security, efficiency, and sustainability.

5.2.1 Examples of smart contracts

Here are some examples of smart contract clauses by sector so you can get an idea of how they work:

- Financial sector: If the user account balance falls below the minimum threshold of €500, the smart contract will automatically initiate a transfer from the linked account to replenish the balance up to the set threshold.
- Real estate: When the agreed payment for the property has been received in full, the smart contract will automatically transfer ownership of the property to the buyer so that the transfer is recorded on the blockchain.
- Healthcare: If the patient grants access to his/her medical data to a specific provider, the smart contract will allow access to the medical information stored on the blockchain for the authorized period.
- Logistics sector: The smart contract will release payment to the supplier once the system confirms that the merchandize has been delivered to the final destination and has been electronically signed by the recipient.
- Energy sector: The smart contract will automatically calculate and generate an invoice at the end of each energy consumption cycle based on consumption data recorded in real-time on the blockchain [27].

5.3 BLOCKCHAIN

To understand smart contracts, it is necessary to know how they work. Blockchain is a distributed data platform replicated and shared among network members. This concept is deduced from what was outlined by Satoshi Nakamoto in the whitepaper [28] that gave life to Bitcoin, one of the most famous cryptocurrencies in the world. In this paper, blockchain was proposed as the solution to the problem of double currency spending. Satoshi Nakamoto is the pseudonym used by the person or group of people to publish the whitepaper on the metzdowd.com cryptography mailing list in 2008. This paper describes a P2P system that makes the trade and use of crypto assets feasible. While Bitcoin and crypto assets, in general, pose challenges and deserve further consideration, what interests us in this essay are smart contracts and other applications of this technology; a blockchain can stand on its own without the need to be linked to cryptocurrencies. Starting from the previous note, to make the blockchain more understandable, let's imagine it as a digital ledger whose records are grouped into blocks in which a "footprint" containing a timestamp remains. Each block is identified by its cryptographic hash. In the new block, reference is made to the hash of the previous block; this mechanism establishes a link between the blocks, thus creating a chain with information; any node or computer with access to this systematic and linked list of blocks can read it and therefore know what is the status of the data exchanged on the network, but without knowing the information contained in them [29]. Cryptographic hash functions are used to ensure the integrity of messages; this allows us to be sure that some communications or files were not altered in any way, and the hashes created before and after the transmission of the data can be examined. If the hashes are the same, there has been no alteration. Blockchain is a revolutionary system that protects itself thanks to the distribution and consensus structure given by the participants and not by a centralized entity. This system provides high security against hacks or computer crimes that put information at risk. Another advantage of this system is that, even if the network were to go down, as long as only one of those computers or nodes did not go down, the information would never be lost. In this sense, falsifying an entry in the blockchain would be equivalent to getting more than half of the participants to agree to lie about the details contained in this chain, all simultaneously and without having the possibility of coordinating beforehand.

What does a blockchain essentially contain?

- Information: Regarding value transfers, the block contains Sender - Receiver - Date - Time: minute and second - Amount and, generally, the sequence that said transaction has followed.
- Hash: This is the number that identifies the block. Each block contains a unique and unrepeatable number plus the number of the previous block. This allows the chain to be formed. This "identification" number will depend on the information it contains. If the information changes, the number will also change, the chain will be broken, and all participants in the network will realize the attempted fraud.

It is essential to mention that in public blockchains (like Bitcoin), consensus can be expensive due to the "miners" issue; however, in private blockchains where participants are on an "allowlist," expensive consensus mechanisms like proof of work are not needed; there is no risk of a Sybil attack, thus eliminating the need for an economic incentive for mining and enabling a more comprehensive range of consensus protocols to choose from [30].

Blockchain technology becomes the most appropriate option to achieve the above purposes of its distributed structure; it allows information to be certified and traceable at all times without the need for expensive platforms, thus having complete security and certainty of the information contained in the digital file, it is also an initiative that is related to the Digital Government policy and the digitalization and public innovation strategy, which seeks and facilitates the promotion of modernization, real access and equity in what has to do with justice services.

Blockchain and smart contracts can work through different types of networks, and it is also made up of a series of elements that are necessary for their proper functioning, which will be described below.

There are different types of blockchain networks, which can be classified into public, private, semi-private networks, and those built through a consortium, each with different characteristics. These characteristics and the operation of each of the different networks are described as follows:

- **Pubic:** The reading of the data is not restricted. Any user can read and generate transactions.
- **Private:** This network focuses on all data access only for users in its own network.
- **Semi-private:** In this case, the operation is carried out by a single entity that grants access to users who meet the requirements.
- **Consortia:** The control is held by a pre-selected group, and access can be public or restricted.

Hasselgren et al. [31] reviewed the current use of the different types of blockchain networks in the medical area and found that the consortium-type network is the most used in health.

For blockchain and smart contracts to be executed in different types of networks, a series of essential elements is needed for their operation.

Limitations and benefits of the operation of blockchain and smart contracts [32]: Some limitations can be identified in the development and operation of projects of this type; the most important is the shortage of skilled labor or blockchain developers; the time required to create a DApp is very long, on average 1800 hours per person of work for an experienced developer, the value of labour is costly. Consequently, all these factors make it difficult to produce this new technology smoothly.

Globalization and the appearance of the internet, associated with the dizzying speed of the development of new technologies, have led society in the constant search for its implementation to generate common benefits. Here, the need for exchanging information and realizing safe, transparent, and reliable transactions of value arises. So, blockchain appears as an important option that allows for solving a large part of these situations.

It can also be seen as a solution not only for exchanging information but for realizing reliable value transactions that can be easily audited. One of the most important qualities is its capacity for decentralization and automaticity.

One of the main benefits of this technology for the health sector is based on the fact that it eliminates the need for centralization of the different processes, opening the doors to the possibility of two or more parties carrying out transactions in an environment without a centralized authority, which improves the problem of the single point. It accelerates the speed of transactions and lowers their cost. Instead of a single authority, a consensus mechanism is used.

Blockchain shows a new variety of possibilities for different projects that allow value to be transferred between users, certify ownership, trace products, or manage the exploitation of digital assets. There are a few advantages that characterize the blockchain:

- The exchange without the intervention of third parties, reducing it to an exchange of assets between two parties, reducing risks considerably.
- Inviolability is invulnerable to malicious attacks, since it lacks weak points, due to the use of decentralized networks.
- It is transparent because the data is available globally, and can be verified and transmitted in real-time.
- The user is in control; they can control the transactions and their information.
- Immutability: transactions cannot be modified or deleted.
- Transactions are efficient, due to security, speed, and effectiveness, which reduces unnecessary intermediary expenses and costs with less monitoring and control.

The health sector could benefit greatly from smart contracts; some examples of their use are identity management to control access to medical records, traceability of pharmaceutical products, contracting health insurance, and development of clinical trials, among others.

5.4 THE IOT

The IoT refers to a global network of identifiable, interoperable, and interconnected devices using the Internet as the primary communication framework. Such devices map the real and digital worlds using front-end computing devices and back-end services [33]. The front-end device can be an embedded computer system with sensors, such as temperature sensors, radio frequency identification (RFID) tags and readers, wearable devices, cameras, smoke detectors, etc. The back-end system, which is software, integrates, processes, and analyzes the information collected and can also return the analyzed results to users.

The advancement in the use of smartphones has further expanded the possibilities of IoT applications in everyday life, since, in addition to being equipped with more powerful processors and greater storage capacity, they facilitate their users' interaction with other IoT objects in the environment.

In addition to individual use, cyber-physical systems (CPS) have contributed to advancing the IoT in industrial environments. Intelligent factories emerged from the union between CBS and IoT, capable of closely monitoring the information produced by heterogeneous embedded devices through software.

It is noted that, by their very nature, IoT applications have unique characteristics that directly influence blockchain use. The following are the main characteristics listed and briefly described.

- Distribution: Devices are deployed in geographically distributed locations. Information computing and service provision can be located at the edges of the network, very close to devices, or on centralized cloud platforms.
- Computational capacity: IoT devices range from small embedded sensors with minimal features, such as processing power, storage, and battery life, to advanced high-capacity servers, such as fog computing. Resource restriction can result in the vulnerability of devices to malicious attacks.
- Large number of devices and data: the amount of data produced by smart devices is growing exponentially due to the expansion of IoT systems.
- Heterogeneity: IoT systems involve multiple devices, each with differences in software and hardware, and following different communication standards and protocols. Consequently, various data types exist (structured, semi-structured, and unstructured). Heterogeneity is also the root of other challenges, such as interoperability, privacy, and security.
- Dynamic: IoT systems are dynamic, as devices are connected and disconnected from the network at any time. Devices, software, and networks can be flawed or compromised. Device volatility is widespread in IoT environments [34].

5.5 CHALLENGES AND OPPORTUNITIES IN INTEGRATING IOT AND BLOCKCHAIN TECHNOLOGIES

The integration of IoT and blockchain presents both challenges and opportunities. Challenges include scalability issues due to high transaction processing times, security vulnerabilities in resource-constrained IoT devices, and high energy consumption of blockchain consensus mechanisms. Additionally, interoperability between different IoT and blockchain platforms remains a concern. Opportunities include enhanced security through decentralized authentication, improved data integrity via tamper-proof ledgers, and increased automation with smart contracts. Blockchain also strengthens IoT-based supply chains and smart city applications by ensuring transparency and trust. Addressing these challenges will enable a more secure, efficient, and scalable IoT blockchain ecosystem. Integrating IoT and blockchain offers promising applications but presents significant challenges. Adhikari and Ramkumar [35] explore how blockchain enhances IoT security, data integrity, and automation in supply chains and smart cities. Albakri [36] highlights opportunities in decentralized identity management and secure communication but emphasizes concerns over scalability and energy consumption. Zafar et al. [37] discuss interoperability challenges, privacy issues, and computational constraints while proposing potential solutions like lightweight consensus mechanisms. These studies underscore blockchain's potential to secure IoT networks, though addressing technical and scalability limitations is crucial for widespread adoption.

5.6 IOT DATA MARKET BASED ON BLOCKCHAIN TECHNOLOGIES

Access to data flows generated by IoT infrastructures can be circumscribed within cloud computing models that consider that any information is available remotely through web technologies, and more specifically, to the sensing as a service (SaaS) model through which applications acquire the necessary context information using these service-oriented architectures.

Focusing on the smart city ecosystem, Anusha et al. [38] propose a functional architecture for these scenarios based on three actors: data generators, service providers, and consumers of data and services. The BIDM concept can be extrapolated to this approach since it considers the regulated trade of data between the owners of the IoT infrastructure, those who expose the aggregated context information using services, and those who make intelligent use of the available information.

Initially, raw, unprocessed sensor information was accessible through centralized systems in the cloud. This data market model, which could arouse certain reservations in terms of privacy for data producers who were not, in turn, the managers of these centralized systems, evolved towards P2P solutions [38], whose distributed nature minimizes the probability of total system failure by avoiding the potential single point of failure of previous solutions. However, although distributed, they have yet to respond to the scalability problems in providing a globally shared and interoperable registry between IoT platforms, an aspect considered by the BIDM proposal presented in this chapter.

Another aspect to consider in deploying distributed data markets is the need for a trust management model to validate transactions without the need for a central trusted entity. Alghofaili and Rassam [39] outline the metrics and properties (security, reliability, availability, precision, etc.)

that the metadata associated with the context information generated in the IoT field must include to provide the necessary trust for its use. These can be easily added to the data model considered in BIDM. Moreover, BIDM extends trust beyond the data, considering also the producers and consumers themselves.

Focusing on more recent solutions in which distributed support is based on blockchain technologies; there is a growing interest in solutions that explore integrating these with IoT. Most of them focus on specific verticals (autonomous vehicles, health, food traceability, etc.) instead of looking for a transversal solution like the one discussed in this chapter.

However, there are various architectures to support data markets with approaches similar to the one described here. The Marketplace for Aealtime (or Real-Time) Human Sensing Data (MARSA) platform [40], while sharing an objective in that it categorizes producers and their linked data. It establishes the mechanisms to trade with them, considers using a central entity that acts as a trusted intermediary between the parties involved in a transaction. The proposal described by Misura also relies on an agent that monitors the transactions, but in this case, the exchange is direct between the parties. Therefore, it is a hybrid solution that is in line with BIDM. However, it should be noted that in the case of BIDM, the central agent acts only in the storage of data, basing the rest of the transactions on a distributed system on blockchain. Additionally, BIDM supports continuous data transmission, unlike Misura's solution, which is based on the request-response model.

5.6.1 Architecture of IoT data market and blockchain

Finally, it is worth highlighting the work of Bhatt et al. [41], who describe a market as a set of smart contracts running on an Ethereum blockchain network, which accesses data stored in Swarm [42], a distributed information storage environment. Although the solution has excellent similarities with BIDM, it differs in storage. The BIDM, in its objective of achieving interoperability and federation between IoT environments, relies on a standard and open-source solution widely adopted by the IoT community to develop intelligent ecosystems such as FIWARE [43].

One of the main objectives of the platform presented in this chapter is to integrate blockchain technology into any current or future IoT ecosystem and offer mechanisms to manage data trading securely and reliably. The inherent properties of blockchain are exploited to allow traceability of data and operations, thus meeting the following fundamental design conditions:

- The measurements or data are generated and exported by trusted entities.
- Payment for access to information is made at a specific time.

- Buyers receive the acquired information.
- Only authorized entities and users can access the information securely.

This results in total transparency in the system's operation for consumers or clients and producers or IoT infrastructures.

A. Functional architecture: Figure 5.1 shows the functional architecture of the BIDM and the relationships between the elements that make it up. Several functional groups are identified: on the one hand, the platform management subsystem and the data persistence subsystem and, on the other, the external entities that will be linked to the data providers or consumers. It has been decided to include the elements that make up the blockchain network among the components of the management subsystem since all the intelligence of the market operations has this technology as its central core. The blockchain network on which the BIDM is based, whether public or private, is based on the proof of authority (PoA) consensus protocol and supports the definition of smart contracts. This pair reduces the possible implementations to networks operating with Ethereum-based technology, whether Ethereum itself or others like Quorum. PoA uses the identity and reputation of the validators to guarantee the proper functioning, transparency, and reliability of operations within the network. Considering the objective of supporting the data market within a federation of IoT infrastructures, it is estimated that the natural option to support BIDM is the use of this consensus method together with the deployment of a permissioned blockchain network, where all federated IoT environments have the same weight and, therefore, the same degree of trust. Additionally, PoA, compared to other consensus protocols such as proof-of-work (PoW), is better suited to the asynchronous nature of the IoT since it allows the periodicity of the mined blocks to be set based on the maximum and minimum data publication rates, thus minimizing the number of empty blocks stored in the blockchain.

Going deeper into the main functional components, we find, on the one hand, the counter or showcase as a point of interconnection with the outside and the data persistence subsystem:

- The counter or showcase, considered an I/O interface, is the element through which the measurements sent by the data producers are collected, advertised to potential consumers, and served at the latter's request. It also redirects the information to the persistence subsystem and indexes and references it within the blockchain to facilitate its subsequent search and access. This component can be divided into an input proxy or application programming interface (API) that processes the information received and a blockchain node

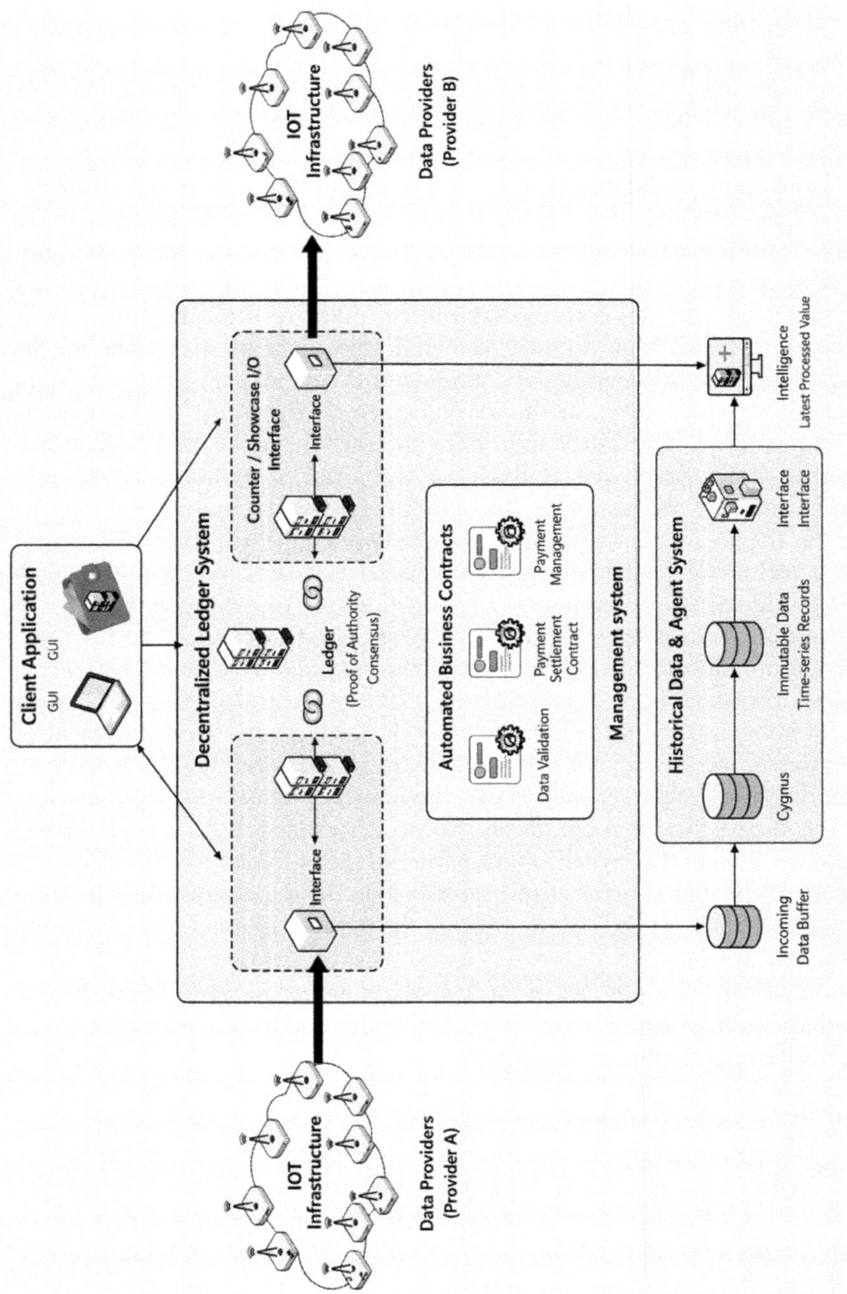

Figure 5.1 Functional architecture of the solution [44].

as an entry point to the network and an enabling element of the interaction with the smart contracts executed in it. There may be instances for each provider that will act as a portfolio. The availability of different entry points to the blockchain network reduces potential scalability problems. The visual interface can be adapted in each instance, although it will enable global access to information.
- The data persistence subsystem is the information store, supported by generic enablers of the FIWARE ecosystem, such as Orion and Cygnus. It also includes a database indexed by the block hash and accessible through a web agent. Using standard and interoperable data models will allow infrastructure sharing between different providers or federating their own.

B. Procedures: In addition to these elements, three intelligent contracts are deployed that provide trust in the BIDM during access control, data storage, and pay-per-use operations. The first limits the registration of measurements or data on the platform to trusted sensors or producers in the same way that it restricts access to information to duly registered and validated clients. The registration of these users, producers, and consumers is done through a user management procedure whose credentials, linked to their Ethereum accounts and digital certificates, are integrated with the user database accessed by the access control contract. The contract related to data storage manages how information is stored in the blockchain. The data model used includes, in addition to the measurement hash, as an indexing parameter and guarantee of integrity, an absolute reference to its location in the data persistence subsystem so that it can be recovered in whole or in part as agreed in the purchase transaction. In addition, a series of descriptive metadata of the measurement are incorporated, which will help advertise it in the market. In the implementation that has been carried out, the information contained in this metadata refers to the physical phenomenon measured in the observation, the sensor from which it comes and the price assigned to the measurement. Still, from a design point of view, it could be extended to include the measurement's quality characteristics, the producer's reputation, etc. Finally, the payment contract manages the measurement acquisition, avoiding duplicate payments and guaranteeing the receipt according to the purchase conditions. Below, the interaction between them is detailed to respond to the needs of authentication and access to information. Only previously authorized sensors or trusted providers can interact with the platform to ensure the origin of the stored data. The authorization process consists of associating a unique identifier to these information sources. The association process can be carried out in multiple ways, including delegating trust to the data market. In the case described in this chapter, it has been decided to generate the credentials offline and record them securely on the devices; that is, the BIDM's particular manager generates an identifier and a unique password for each device. Additionally, the BIDM's public key or digital certificate is provided. Requests for registration of measurements sent by suppliers must, therefore, include, in addition to the measurement itself in NGSIv2 format [45], the credentials that allow the identity of the supplier to be validated. To avoid compromising the confidentiality of the same, both the identifier and the password are encrypted with the BIDM public key.

Upon receipt of the measurement, and once an authorized producer has verified it, the integrity and validity of the information received are checked, and the document is adapted by eliminating non-relevant or redundant information. Subsequently, the hash of the final document is calculated and redirected for storage.

Once the measurement storage has been confirmed by the data persistence subsystem, including a unique and permanent link to the stored document, the hash, reference, and metadata (description, etc.) are injected into the blockchain. The information residing in the blockchain makes it possible to facilitate the search for data based solely on the available data while reducing the size of the data by outsourcing the storage of the measurement data. However, integrity guarantees are maintained by incorporating the reference to the measurement and its summary separately. The blockchain platform and operation itself grant trust in the data set that makes up the document.

Potential clients carry out the search process. Clients are also entities registered and validated in the blockchain network, but unlike IoT providers, they have restricted permissions for reading operations. The search is carried out directly on the blockchain itself. However, as an alternative, a direct search in the persistence environment can be considered through an API generated for this purpose. The result of both searches will be the document's hash that allows referencing the blockchain.

Once the measurement or set of measurements to be acquired is identified, the purchase is carried out, granting the right to access their values. Figure 5.2 shows the development of the process that consumers must follow, involving all smart contracts.

The customer begins the purchase process by initiating a purchase transaction with the intelligent payment contract (step 2 in Figure 5.2), which, supported by the access control contract, guarantees that it is an authorized customer. Among the data included in the purchase transaction, in addition to the customer identifiers, there is also the unique identifier or hash of the measurements to be purchased. Since it is also stored in the blockchain, the contract obtains a unique reference (i.e., URL) for the measurement. Since the persistence subsystem agent offers an interface for accessing information based on RESTful services, the value of the measurement is obtained after the corresponding HTTP request.

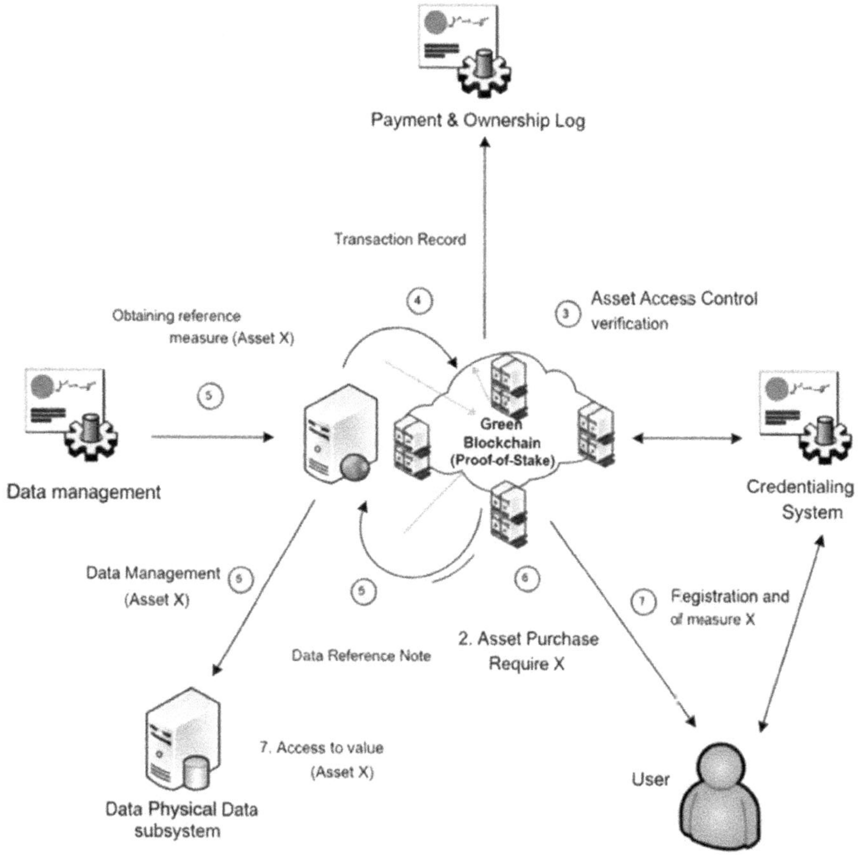

Figure 5.2 Measurement access process [46].

The last phase of the process guarantees that the payment process is only completed when the buyer accesses the data. To do this, the data is conditioned to ensure that only the purchaser can access it before proceeding to its delivery.

This way, the measurement is encrypted using a symmetric encryption algorithm (i.e., advanced encryption standard [AES]) using a sufficiently robust randomly generated key. This key is provided to the client by encrypting it with the public key included in its certificate. In this way, it is guaranteed that only the said client can decrypt the key and, therefore, the measurement.

Additionally, so that the BIDM administrator can access the measurement in the event of a dispute, the key is also encrypted with the public key linked to the BIDM. In this way, if the client reports that the measurement does not correspond to what was purchased, the administrator can check what values were sent to him.

All this information, the keys, and the encrypted measurement are included in a transaction between the BIDM and the client, which is recorded in the blockchain network. Given the public nature of the information in the blockchain blocks, any user with authorized credentials can access this information linked to the purchase. However, only the customer who made the purchase can access the measurement since it is stored in the information persistence subsystem protected according to the previous procedure.

At this point, the payment transaction is concluded and reflected with the corresponding entry in the blockchain. The process described involves storing the measurement in an encrypted blockchain. It may seem inconsistent with the initial premise that storing the measurements outside the blockchain is required in the data persistence component. However, the volume of data suppliers generate far exceeds the purchases made. This is why this methodology has been chosen, since it is considered that the benefits in terms of guaranteed availability compensate for the additional storage space needs.

As an additional process, the client can check that the measurement obtained matches the one requested, thanks to the summary in the blockchain when publishing the measurement.

To conclude, the described processes allow for the complete traceability of the life cycle of data generated by any IoT supplier, both in the generation process itself and in as many acquisition processes as consumers are interested in said observation.

5.6.2 Proof of concept

A. Development and deployment environment: To validate the solution described in this chapter, all the functional components of the BIDM have been developed and integrated into a proof-of-concept implementation. The deployment has been carried out in a virtual environment using Docker containers to facilitate the replicability of the system.

The decision has been made to use a private blockchain network based on Ethereum Clique [46], configured to use the available PoA consensus protocol. The network deployed for validation consists of four nodes, two acting as validators and the other two as entry points to the network (without validation permissions) for a provider and a consumer wallet of context information, respectively. A fifth node is also deployed as an initializer node to support the discovery of the distributed network configuration. Although it may not be considered indispensable, the presence of this node significantly reduces and facilitates the node discovery process in large networks, which is why it is included as support for scalability.

The smart contracts have been developed in Solidity and deployed directly on the blockchain network, thus being available from the start of the network. Finally, for the persistent storage of the context information collected from the IoT infrastructures, the FIWARE enablers Orion Context Broker and Cygnus have been used, both deployed using Docker containers.

The interface of the counter or showcase and the customer portfolio has been created using web technologies, the web service logic in Nodejs and the user interface in HTML and JavaScript.

B. Integration and validation in actual IoT infrastructure: The validation of the implementation carried out is within the IoT infrastructure available in Bangalore (India) and is managed within the scope of the Smart Bangalore project.

As shown in Figure 5.3, the BIDM, for which only one instance of the data showcase has been deployed, subscribes to the measurements generated by the sensors available in Bangalore and is exported through an Orion Context Broker provided for this purpose. Therefore, the Smart Bangalore infrastructure is an authorized trusted provider in the BIDM. The information collected by the BIDM, in NGSIv2 format, is processed and stored in the blockchain associated with the BIDM and in the persistence environment as described, so that any client can request its acquisition. Table 5.1 shows an example of how the measurements are stored in the blockchain. By reading the blockchain, the consumer of context information, in addition to the hash, has access to the descriptive information of said measure (in the

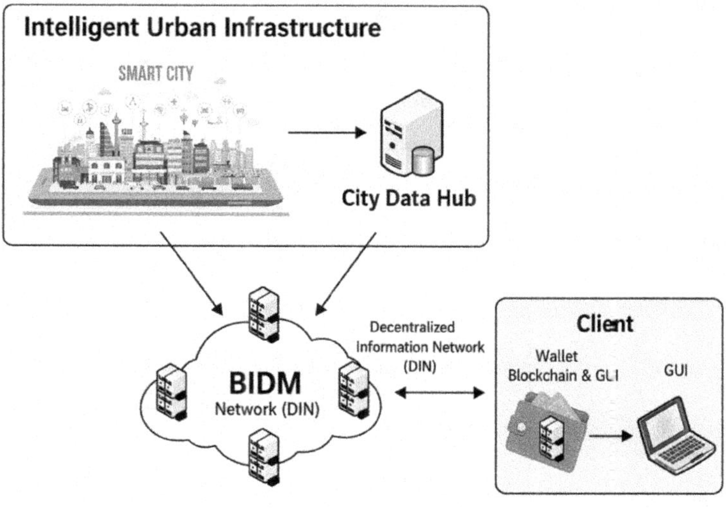

Figure 5.3 Integration of the BIDM in the Smart Bangalore infrastructure [48].

Table 5.1 Data available on the blockchain

		Data Available *Latest Measurements*		
Date	*Topic*	*Transaction Hash*	*Measurement Hash*	*Price*
22-01-2024	iot:smartsantander:u7jcfat5191 [temperature:ambient]	0x5a04e014e9f2d33c361e84d92584c0 2aab 4fd6a2c55c5a203cfd41eb7933 355f	0xceec922bf7720adc83e69a6260b 70222b0f1fde93d5eb5e407e4d3bdd 6ab75b6	1
26-05-2024	iot:smartsantander:u7jcfat5191 [temperature:ambient]	0x449e4eb1071a158ccc9f1b25c044a4 b6737 b5d6dd7a13678be58358b895b910	0xf0284080596759f996d67845 7b652f2af5a9f23e493c4a136d102e37 24f72045	1
25-04-2024	iot:smartsantander:u7jcfat5191 [temperature:ambient]	0x236a8b53ac2f0a8028e77f3809 efc6809394 ca82c43729aa300121ffb9442ec	0x721abe006bf47aa38e76b2f 10a88671 1b19d8d83799db104dd771 2344125df	1

Figure 5.4 User wallet user interface [49].

case of the implementation carried out, this information was the identifier of the sensor that generated the measure, a timestamp, and the price set for its acquisition). An interested and authorized client can purchase their virtual wallet.

Figure 5.4 shows the interface of a simple wallet, implemented as a web page from which the blockchain can be explored, and the information included in the Counter/Showcase can be seen (i.e., the information contained in the measurement publication blocks). Through it, it was possible to obtain the summary of the desired measure and, with it, launch a purchase request, resulting in the encrypted measure and the corresponding key to decrypt it. What is implemented in this proof of concept as a web from which measures are selected and purchases are made, in a typical use case, could be a mobile application that offers a tourist guide service in the city and among its services is the recommendation of routes. To go from one place to another in the same. The application will act as a client and will have its virtual wallet. When the user requests a recommendation for a trip, the application will buy from the BIDM the information regarding the arrival times of urban buses at the nearest stop, the bicycles available at the public rental service stations closest to the origin and destination, and the number of parking spaces in the vicinity of the destination, to offer different travel alternatives to the user. On the other hand, the BIDM publishes and allows access to the information once the payment has been completed and the imposed conditions of use are met. The BIDM guarantees, in a certain way, the quality of the information and its origin. In this situation, all participants in the service chain obtain a benefit within an ecosystem of mutual trust. The remuneration received from the sale of the data will allow the provider to improve and expand its infrastructure, which will redound to the benefit of the application developer, who can offer it in better conditions to a more significant number of users.

Figure 5.5 shows a screenshot of the developed wallet application, which shows the information available at the end of a purchase, including the transaction hash that incorporates the symmetric encryption key to the blockchain (KeyTxHash) and the transaction hash linked to the transfer of the encrypted requested measurement (DataTxHash), in addition to the hash of the information event at the end of the purchase (EventTxHash). Although the user could access the measurement value through these references, the application directly shows the measurement in plain text to facilitate the client's operation. In the example, a measurement is provided from a sensor that monitors the traffic flow at a specific position.

5.7 CONCLUSION

IoT and blockchain are rapidly growing technological advances that give rise to new business opportunities, with the potential to revolutionize the most diverse application areas. For this reason, they have been growing in terms of research and investments by industries. The chapter showed the aspects that need to be improved in IoT systems and what benefits joining with blockchain technology can add. In addition, some proposed architectures were presented, which are models that modify conventional architectures with the primary objective of increasing the scalability and performance of IoT applications. Some studies have already proposed specific applications related to different domains of application, such as health, energy, smart cities, logistics, and agriculture. Although it is not possible to find a universal solution applicable to all scenarios, these proposals are essential to confirm, in practice, the foreseen limitations and, above all, to foster research aimed at the union between blockchain and IoT, an area that is as promising as it is emerging.

This chapter presents a platform that uses blockchain technology to enable a decentralized IoT data marketplace. Through the BIDM, IoT infrastructures and context data-consuming applications can exchange information reliably and transparently.

KeyTxHash: 0x852e5a2471996afe0240f02b5bb3e2e085bf998480617c0b24e42b2c073fe416

DataTxHash: 0xa9ee79edc878e96be46060aed923f07367bd2924e8e239a8b1848cf9b0a2f073

EventTxHash: 0x8de09b749688386aed5634060e4953d6379aa46be646fc2e624f9b919e49467c

Data: {"recvTimeTs":"1601372025035", "recvTime":"2023-09-29 09:33:45.35","fiware
ServicePath":"/trafficflowobserv ","entityId":"urn:ngsi-Id: Traffic FlowObserved:bengalore:traffic:flow:1018",
"entityType":"TrafficFlowObserved","attr Name":"attributes","attrType":"Object","attrValue":"{\"dateModified\":\"2023-0929T09:32:00.00Z\",\"
dateObserved \":\"2023-09-29T09:32:00.00Z\",\"intensity\":840,\"laneld\":0,\"location\":{\"coordinates\":[3.8087975,
43.4584602],\"type\":\"Point\"},\"occupancy\":0.1,\"roadLoad\":36,\"sensorID\":{\"type\":\"String\",\"value\":\"iotsmartbengalore1\"}}","attrMd":
"[{\"name\":\"hash\",\"type\":\"String\",\"value\":\"b31eb92211b0a89b42d13ccbabd30aee5a3818a4f6f56 9dbc99e19c6bcfecd4b\"}]"}

Figure 5.5 Example of a customer-acquired measure [49].

The solution proposed and described in the chapter combines IoT ecosystems based on the blocks defined within the scope of Connecting Europe's Facilities (CEF), such as the Orion Context Broker, with blockchain technology to create a novel environment that allows the monetization of data flows with traceability guarantees throughout the data lifecycle. Thanks to the authentication and authorization mechanisms and the inherent properties of blockchain technology, the data's veracity, reliability, and quality can be guaranteed.

The solution addresses potential scalability problems resulting from the high and unsustainable growth of the blockchain site. To do this, an off-chain data storage strategy is applied. Instead of storing all IoT data in the blockchain itself, it is stored in a specialized environment integrated into IoT infrastructures, maintaining in the blockchain only an immutable reference to the transactions carried out with said data (i.e., registration and purchase-sale). Additionally, the designed solution supports secure and guaranteed access to the information only to those users who, after paying for it, have acquired it. The combination of the blockchain, the execution of smart contracts, and additional encryption mechanisms makes this possible.

Although the deployed solution has demonstrated the viability of a reliable and traceable data market that combines standard IoT platform solutions (i.e., open functional enablers of the FIWARE ecosystem) and blockchain networks, capable of being integrated into a natural environment such as the ecosystem of the city of Bangalore, there are still several improvements to be applied. In this sense, future work is planned to include an exhaustive performance analysis and study of the system's scalability, emphasizing not only aspects related to data security and storage capacity but also quality of service and user experience as the volume of data increases. In addition, the search for new, more robust, and user-friendly mechanisms to manage the life cycle of stored information from its generation to its acquisition and use by the end client is proposed. Finally, the extension of BIDM is proposed to support information consumers with real-time needs, for which it is necessary to enable asynchronous access (i.e., based on subscriptions) to the measurements.

Finally, this work is expected to motivate and contribute positively to the Computing Applied to various communities. Here, a general and comprehensive view of a set of technologies was presented, which do not have trivial methods to operate together.

Challenges and future: Despite its numerous advantages, blockchain integration with IoT faces significant challenges. Scalability is one of the main issues, as current blockchains may not handle the massive volume of transactions that a fully developed IoT ecosystem will generate. Furthermore, power and processing limitations in IoT devices can make it challenging to implement the advanced cryptography needed to participate in a blockchain efficiently.

However, research into technologies such as next-generation blockchains, such as proof-of-stake (PoS) blockchains or second-layer networks, could mitigate these challenges. Furthermore, developing more robust and energy-efficient IoTs will facilitate integration.

REFERENCES

[1] Bodo, B., Gervais, D., & Quintais, J. (2018). Blockchain and smart contracts: The missing link in copyright licensing? *International Journal of Law and Information Technology, 26*(3), 311–336. https://doi.org/10.1093/ijlit/eay014

[2] Malik, H., Anees, T., Faheem, M., Chaudhry, M. U., Ali, A., & Asghar, M. (2023). Blockchain and Internet of Things in smart cities and drug supply management: Open issues, opportunities, and future directions. *Internet of Things*, *23*, 100860. https://doi.org/10.1016/j.iot.2023.100860

[3] Mishra, S. (2023). Blockchain and machine learning-based hybrid IDS to protect smart networks and preserve privacy. *Electronics, 12*(16), 3524. https://doi.org/10.3390/electronics12163524

[4] Singh, S., Rosak-Szyrocka, J., & Tamándl, L. (2023). Development, service-oriented architecture, and security of blockchain technology for Industry 4.0 IoT application. *HighTech and Innovation Journal, 4*(1), 134–156. https://doi.org/10.28991/HIJ-2023-04-01-010

[5] Raja Santhi, A., & Muthuswamy, P. (2022). Influence of blockchain technology in manufacturing supply chain and logistics. *Logistics, 6*(1), 15. https://doi.org/10.3390/logistics6010015

[6] Dong, S., Abbas, K., Li, M., & Kamruzzaman, J. (2023). Blockchain technology and application: An overview. *PeerJ Computer Science, 9*, e1705. https://doi.org/10.7717/peerj-cs.1705

[7] Christidis, J., Karkazis, P. A., Papadopoulos, P., & Leligou, H. C. (Nelly). (2022). Decentralized blockchain-based IoT data marketplaces. *Journal of Sensor and Actuator Networks, 11*(3), 39. https://doi.org/10.3390/jsan11030039

[8] Sober, M., Scaffino, G., Schulte, S., & Kanhere, S. (2022). A blockchain-based IoT data marketplace. *Cluster Computing, 26*(1), 1–23. https://doi.org/10.1007/s10586-022-03745-6

[9] Taherdoost, H. (2023). Smart contracts in blockchain technology: A critical review. *Information, 14*(2), 117. https://doi.org/10.3390/info14020117

[10] Negara, E. S., Hidayanto, A. N., Andryani, R., & Syaputra, R. (2021). Survey of smart contract framework and its application. *Information, 12*(7), 257. https://doi.org/10.3390/info12070257

[11] Choudhary, V., Guha, P., Pau, G., & Mishra, S. (2025). An overview of smart agriculture using internet of things (IoT) and web services. *Environmental and Sustainability Indicators, 26*, 100607. https://doi.org/10.1016/j.indic.2025.100607

[12] Ayaz, M., Ammad-Uddin, M., Sharif, Z., Mansour, A., & Aggoune, E.-H. M. (2019). Internet-of-Things (IoT)-based smart agriculture: Toward making the fields talk. *IEEE Access, 7*, 129551–129583. https://doi.org/10.1109/ACCESS.2019.2932609

[13] Yang, Y., Lin, M., Lin, Y., Zhang, C., & Wu, C. (2025). A survey of blockchain applications for management in agriculture and livestock Internet of Things. *Future Internet, 17*(1), 40. https://doi.org/10.3390/fi17010040

[14] Puthenveettil, N. R., & Sappati, P. K. (2024). A review of smart contract adoption in agriculture and food industry. *Computers and Electronics in Agriculture, 223*, 109061. https://doi.org/10.1016/j.compag.2024.109061

[15] Peng, X., Zhao, Z., Wang, X., Li, H., Xu, J., & Zhang, X. (2023). A review on blockchain smart contracts in the agri-food industry: Current state, application challenges and future trends. *Computers and Electronics in Agriculture, 208*, 107776. https://doi.org/10.1016/j.compag.2023.107776

[16] Pranto, T. H., Noman, A. A., Mahmud, A., & Haque, A. B. (2021). Blockchain and smart contract for IoT-enabled smart agriculture. *PeerJ Computer Science, 7*, e407. https://doi.org/10.7717/peerj-cs.407

[17] Marzougui, F., & Elleuch, M. (2025). Blockchain and IoT in smart agriculture: Analysis, opportunities, challenges, and future research directions. *Journal of Information Assurance and Security, 19*, 104–119. https://doi.org/10.2478/ias-2024-0008

[18] Duguma, A. L., & Bai, X. (2025). How the internet of things technology improves agricultural efficiency. *Artificial Intelligence Review, 58*, 63. https://doi.org/10.1007/s10462-024-11046-0

[19] Dhanaraju, M., Chenniappan, P., Ramalingam, K., Pazhanivelan, S., & Kaliaperumal, R. (2022). Smart farming: Internet of Things (IoT)-based sustainable agriculture. *Agriculture, 12*(10), 1745. https://doi.org/10.3390/agriculture12101745

[20] Malik, H., Anees, T., Faheem, M., Chaudhry, M. U., Ali, A., & Asghar, M. N. (2023). Blockchain and Internet of Things in smart cities and drug supply management: Open issues, opportunities, and future directions. *Internet of Things, 23*, 100860. https://doi.org/10.1016/j.iot.2023.100860

[21] Ullah, Z., Naeem, M., Coronato, A., Ribino, P., & De Pietro, G. (2023). Blockchain applications in sustainable smart cities. *Sustainable Cities and Society, 97*, 104697. https://doi.org/10.1016/j.scs.2023.104697

[22] Imad, B., Anass, S., Mounir, A., & Khalid, C. (2024). Blockchain-based smart contract to enhance security in smart city. *2024 11th International Conference on Wireless Networks and Mobile Communications (WINCOM)*, Leeds, United Kingdom, 1–6. https://doi.org/10.1109/WINCOM62286.2024.10657758

[23] Rejeb, A., Rejeb, K., Simske, S. J., & Keogh, J. G. (2022). Blockchain technology in the smart city: A bibliometric review. *Quality & Quantity, 56*(5), 2875–2906. https://doi.org/10.1007/s11135-021-01251-2

[24] Biasin, M., & Delle Foglie, A. (2024). Blockchain and smart cities for inclusive and sustainable communities: A bibliometric and systematic literature review. *Sustainability, 16*(15), 6669. https://doi.org/10.3390/su16156669

[25] Margret, M. K., & Julie, E. G. (2024). Smarter and resilient smart contracts applications for smart cities environment using blockchain technology. *Automatika, 65*(2), 572–583. https://doi.org/10.1080/00051144.2024.2307228

[26] Dritsas, E., & Trigka, M. (2024). Machine learning for blockchain and IoT systems in smart cities: A survey. *Future Internet, 16*(9), 324. https://doi.org/10.3390/fi16090324

[27] Kontos, C., Panagiotakopoulos, T., & Kameas, A. (2024). Applications of blockchain and smart contracts to address challenges of cooperative, connected, and automated mobility. *Sensors, 24*(19), 6273. https://doi.org/10.3390/s24196273

[28] Bayer, D., Haber, S., & Stornetta, W. S. (1993). Improving the efficiency and reliability of digital time-stamping. In *Sequences II* (pp. 329–334). Springer, Berlin, Germany.

[29] Deng, W., Huang, T., & Wang, H. (2023). A review of the key technology in a blockchain building decentralized trust platform. *Mathematics, 11*(1), 101. https://doi.org/10.3390/math11010101

[30] Kõlvart, M., Poola, M., & Rull, A. (2016). Smart contracts. In T. Kerikmäe & A. Rull (Eds.), *The future of law and eTechnologies* (pp. 133–147). Springer, Cham. https://doi.org/10.1007/978-3-319-26896-5_7

[31] Hasselgren, A., Kralevska, K., Gligoroski, D., Pedersen, S. A., & Faxvaag, A. (2020). Blockchain in Healthcare and Health Sciences—A Scoping Review. *International Journal of Medical Informatics, 134*, Art. no. 104040. https://doi.org/10.1016/j.ijmedinf.2019.104040

[32] Habib, G., Sharma, S., Ibrahim, S., Ahmad, I., Qureshi, S., & Ishfaq, M. (2022). Blockchain technology: Benefits, challenges, applications, and integration of blockchain technology with cloud computing. *Future Internet, 14*(11), 341. https://doi.org/10.3390/fi14110341

[33] Mansour, M., Gamal, A., Ahmed, A. I., Said, L. A., Elbaz, A., Herencsar, N., & Soltan, A. (2023). Internet of Things: A comprehensive overview on protocols, architectures, technologies, simulation tools, and future directions. *Energies, 16*(8), 3465. https://doi.org/10.3390/en16083465

[34] Ali, Z., Ali, H., & Badawy, M. (2015). Internet of Things (IoT): Definitions, challenges, and recent research directions. *International Journal of Computer Applications, 128*, 975–8887.

[35] Adhikari, N., & Ramkumar, M. (2023). IoT and blockchain integration: Applications, opportunities, and challenges. *Network, 3*(1), 115–141. https://doi.org/10.3390/network3010006

[36] Albakri, A. O. (2021). Blockchain and the Internet of Things: Opportunities and challenges. *2021 International Conference on Software Engineering & Computer Systems and 4th International Conference on Computational Science and Information Management (ICSECS-ICOCSIM)*, Pekan, Malaysia, 183–188. https://doi.org/10.1109/ICSECS52883.2021.00040

[37] Zafar, S., Bhatti, K., Shabbir, M., Hashmat, F., & Akbar, A. (2021). Integration of blockchain and Internet of Things: Challenges and solutions. *Annals of Telecommunications, 77*, 13–32. https://doi.org/10.1007/s12243-021-00858-8

[38] Anusha, G., Arora, K., Sharma, H., Thota, S. P., Joshi, G. P., & Cho, W. (2024). Control strategies of 15-level modified cascaded H-bridge MLI with solar PV and energy storage system. *Energy Reports, 12*, 2–26. https://doi.org/10.1016/j.egyr.2024.06.003

[39] Alghofaili, Y., & Rassam, M. A. (2022). A trust management model for IoT devices and services based on the multi-criteria decision-making approach and deep long short-term memory technique. *Sensors, 22*(2), 634. https://doi.org/10.3390/s22020634

[40] Cao, T.-D., Pham, T. V., Vu, Q. H., Truong, H. L., Le, D.-H., & Dustdar, S. (2016). MARSA: A marketplace for mobile applications. *ACM Transactions on Internet Technology, 16*(1), 1–21. https://doi.org/10.1145/2883611

[41] Bhatt, B., Sharma, H., Arora, K., Joshi, G. P., & Shrestha, B. (2023). Levy flight-based improved grey wolf optimization: A solution for various engineering problems. *Mathematics, 11*(7), 1745. https://doi.org/10.3390/math11071745

[42] Tron, V. (2016). Swarm alpha public pilot and the basics of Swarm. *Ethereum Blog.*

[43] Cirillo, F., Solmaz, G., Berz, E. L., Bauer, M., Cheng, B., & Kovacs, E. (2020). A standard-based open source IoT platform: FIWARE. *IEEE Internet of Things Magazine, 2*(3), 12–18. https://doi.org/10.1109/IOTM.0001.1800022

[44] Khashan, O. A., & Khafajah, N. M. (2023). Efficient hybrid centralized and blockchain-based authentication architecture for heterogeneous IoT systems. *Journal of King Saud University – Computer and Information Sciences, 35*(2), 726–739. https://doi.org/10.1016/j.jksuci.2023.01.011

[45] Kouloglou, I.-O., Antzoulatos, G., Vosinakis, G., Lombardo, F., Abella, A., Bakratsas, M., Moumtzidou, A., Maltezos, E., Gialampoukidis, I., Ouzounoglou, E., et al. (2024). FIWARE-compatible smart data models for satellite imagery and flood risk assessment to enhance data management. *Information, 15*(5), 257. https://doi.org/10.3390/info15050257

[46] Hang, L., & Kim, D.-H. (2019). Design and implementation of an integrated IoT blockchain platform for sensing data integrity. *Sensors, 19*(10), 2228. https://doi.org/10.3390/s19102228

[47] Oyinloye, D. P., Teh, J. S., Jamil, N., & Alawida, M. (2021). Blockchain consensus: An overview of alternative protocols. *Symmetry, 13*(8), 1363. https://doi.org/10.3390/sym13081363

[48] Sánchez, L., Lanza, J., González, I., Santana, J. R., & Sotres, P. (2023). Development of a blockchain-based marketplace for the Internet of Things data. *ICT Express, 9*(4), 628–634. https://doi.org/10.1016/j.icte.2023.01.002

[49] Lanza, J., González, I., Sánchez, L., Santana, J. R., & Sotres, P. (2021). IoT data market based on blockchain technologies. *Proceedings of the 15th Conference on Telematics Engineering (JITEL 2021)*, A Coruña, Spain, October 27–29, 2021, 47–54.

Chapter 6

Automation and control in smart grids

Sonia Vatta

6.1 INTRODUCTION

Automation and control are key components of smart grids. They enable efficient, robust, and flexible operation of power systems. These systems use sophisticated communication, sensing, and control technologies to observe and regulate electrical power flow in real time. The work by Khan et al. (2018) states that they ensure grid stability and harmonious integration of renewable energy sources.

6.1.1 Evolution of automation in power systems

The evolution of automation in power systems has been a long way from manual operations to automated operations. Earlier power systems used extensive manual operations and relay protection schemes. They provided inherent automation but limited responsiveness and flexibility. The application of supervisory control and data acquisition (SCADA) systems proved as an enhanced development. They provide centralized control and monitoring for better operational efficiency (Zhao & Li, 2019). Subsequently, the use of intelligent electronic devices (IEDs) and communication protocols results in enhanced automation capabilities such as automatic fault detection and system restoration. Recently, the deployment of real-time data analysis, artificial intelligence (AI), and distributed control systems transformed legacy grids into smart grids. Chen & Wang (2020) states that these grids are capable of self-healing, adaptive control, and renewable energy integration. This transformation is a paradigm shift from static, manual processes towards dynamic control systems that are automated and ensure maximum grid resilience and performance. Table 6.1 presents the evolution of automation in power systems, covering key technologies and levels of automation.

DOI: 10.1201/9781003689430-6

Table 6.1 Evolution of automation in power systems

S. No.	*Period*	*Key technologies*	*Automation level*
1.	Pre-1950	Manual controls	Very low
2.	1950–1960s	Telemetry, supervisory control	Basic remote control
3.	1970–1980s	SCADA, digital control centers	Moderate automation
4.	1980–1990s	Microprocessor relays	High-speed protection
5.	1990–2000s	IEDs, PMUs, IEC 61850	Intelligent automation
6.	2000–2015	Smart grid, AMI, DA	Adaptive, distributed
7.	2015-Now	IoT, Edge AI, digital twins	Real-time intelligent automation
8.	Future	AI-autonomous systems	Fully autonomous grids

6.1.2 Automation and control in smart grid vision

Automation and control systems are central to the development of the smart grids. They aim to develop intelligent, flexible, and resilient power grids. These power grids are able to integrate renewable energy and ensure optimal operation. Such types of systems enable real-time collection of data, analysis, and decision-making. Hossain & Muhammad (2018) noted that the grid responds dynamically to dynamic conditions such as fluctuations in supply and demand.

Automation features such as fault detection, isolation, and system restoration (FDIR) in smart grids have an enormous impact on reliability. These features also reduce the time of outages.

The research by Zhang et al. (2019) states that sophisticated control methods, like distributed control and real-time optimization, enable coordinated operation of various grid elements. These elements include distributed energy resources (DERs), energy storage, and demand response (DR) mechanisms. Also, automation and control enable the use of renewable energy sources through the proper control of their intermittency and variability. They support demand-side management, enhance grid stability, and enable the use of smart devices. Based on the research by Li & Wang (2020), these technologies are keys to bringing the vision of the smart grid through the creation of an adaptive energy infrastructure that is capable of meeting future challenges.

6.2 FUNDAMENTALS OF SMART GRID AUTOMATION

Smart grid automation involves the integration of cutting-edge technologies, i.e., sensors, networks, automation software, and analytics into traditional power grids to improve operational efficiency, reliability, and resiliency.

The deployment of automated smart grids makes it possible for real-time monitoring, decentralized energy management, adaptive handling of demand fluctuations, and optimal utilization of renewable resources.

6.2.1 Main drivers of smart grid automation

Reliability, efficiency, and resilience are the main drivers of smart grid automation. These three main drivers are as shown in Figure 6.1.

- *Reliability*

Smart grids improve reliability by employing sensors and automation to track equipment health, anticipate failures, and automatically recover after failure.

This reduces outages and increases energy supply consistency. Real-time data analytics and quick anomaly detection enable proactive maintenance. For instance, predictive maintenance techniques and automated switching ensure a reliable grid, avoiding blackouts.

> "Smart grids utilize state forecasting models, real-time monitoring, and self-repairing features, resulting in a steady and smooth transmission of power and minimizing the risks of disruptions" (Abdullah, 2022).

- *Efficiency*

Real-time load balancing, DR programs, energy loss reduction, and usage-based dynamic adjustment are facilitated to achieve maximum efficiency.

Automation enables utilities to make data-driven decisions, eliminate waste, and optimize generation and distribution assets.

- *Resilience*

Resilience is the ability of the grid to withstand and quickly come back from interruptions such as equipment failure, cyber-attacks, or weather. Self-healing is achieved through automated control of the system, which enables the quick isolation and rerouting under fault conditions to confine the customer impact.

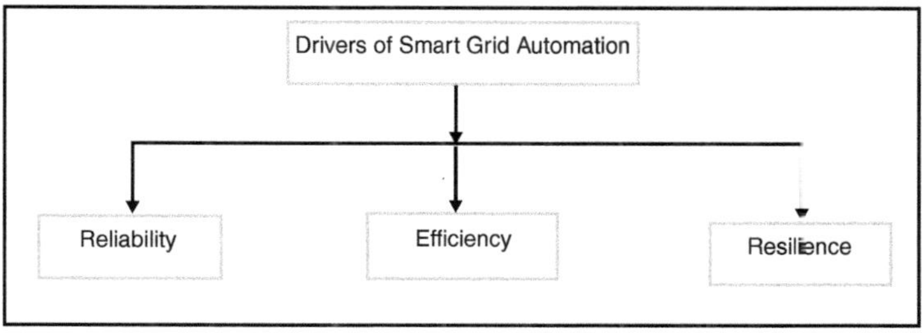

Figure 6.1 Drivers of smart grid automation.

"Resilience improvement is essential to all critical infrastructure networks. Smart grids are made resilient against disturbance and capable of fast fault or outage recovery. Self-healing is achieved because the grid is able to automatically detect a fault, isolate, and reroute power" (Huang et al., 2017).

6.2.2 Hierarchical levels of smart grid automation

Smart grid automation is hierarchically organized on four primary levels: generation, transmission, distribution, and end-user. Each level uses automation technologies unique to its operational requirements that are part of an integrated, responsive, and resilient power system. Figure 6.2 shows hierarchical levels of smart grid automation. It indicates how this hierarchical organization functions to optimize the performance of the grid.

- *Generation level*

At this level, automation entails control systems like automatic generation control (AGC) and energy management systems (EMS). These systems maximize the dispatch of power plants, particularly blending renewables for stability and efficiency. These systems are capable of monitoring real-time grid frequency and voltage. They also regulate generator outputs for equilibrium. For example, to provide the balance between supply and demand, AGC synchronizes the generation units and thus enhances grid stability and reliability.

- *Transmission level*

At the transmission level, automation is associated with the efficient and secure transfer of bulk power across long distances. The prime technologies include wide area monitoring systems (WAMS), SCADA, and phasor measurement units (PMUs). These technologies help to monitor grid conditions

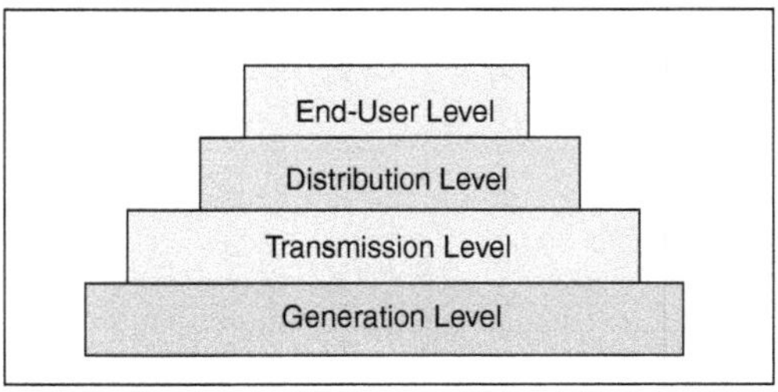

Figure 6.2 Hierarchical levels of smart grid automation.

in real-time, respond instantaneously to disturbances, and make maximum use of lines. Transmission automation results in enhanced grid visibility and control, which in turn enhances system reliability and resilience.

- *Distribution level*

The distribution level experiences considerable automation by way of distribution management systems (DMS), outage management systems (OMS), and advanced metering infrastructure (AMI). These enable utilities to track feeder status, sense faults, manage DER, and dynamically dispatch power. This automation enhances outage response, enables seamless integration of storage and renewables, and enables voltage and load control more efficiently.

- *End-user level*

From the perspective of the end-user, automation enables customers and demand-side management. The important technologies include smart appliances, home energy management systems (HEMS), and smart meters, supporting real-time usage monitoring, flexible load management, and DR programs participation (Mansouri et al., 2025). End-users have the benefits of increased reliability of service and bills that have been optimized.

6.2.3 Communication and interoperability requirements in smart grids

Communication and interoperability are central to the operation, reliability, and security of smart grids. These needs enable a massive universe of devices such as sensors, meters, generators, and control systems. They share information accurately, securely, and in real time, no matter what manufacturer or communications technology they use.

- *Communication requirements*
- Reliability and latency: Networks of smart grids should provide high reliability and low latency to enable real-time monitoring and fast control actions.
- Bandwidth and scalability: Applications, ranging from phasor measurement devices to smart appliances, generate different loads of data. The network must be capable of handling them efficiently and scale to accommodate more grid installations.
- Security and privacy: Security protocols need to be robust enough to protect data integrity and user privacy as millions of devices connect to the grid.
- *Interoperability requirements*

- Device and system interoperability: Devices and systems from various manufacturers need to implement common communication protocols and data formats so that they can work together flawlessly in the grid.
- Standardization: Interoperability relies on the use of standards like IEEE 1547, IEC 61850, IEEE C37.118, and others. These standards specify how devices convey their states, communicate control commands, and synchronize the timing of data.
- Test and certification: Special, testable interoperability profiles and certification schemes are needed to maintain the interoperability. These schemes are helpful to make devices work together as desired (Song et al., 2017).

Each level uses automated technologies unique to its operating conditions. This creates an integrated, responsive, and robust power system. Systems observe real grid voltage and frequency and modulate generator outputs to supply balance.

These systems monitor real-time grid frequency and voltage and set generator outputs to provide balance. AGC, for instance, synchronizes generation units to balance supply and demand and enhance grid stability and reliability.

6.3 CONTROL ARCHITECTURES IN SMART GRIDS

6.3.1 Centralized and decentralized control

- *Centralized control*

In centralized systems, one central controller (e.g., microgrid central controller [MGCC]) collects system-wide information (e.g., voltage, frequency) and calculates control actions for the entire assets in the system. It guarantees coordinated and global optimization, but imposes a heavy computational burden on the central unit and generates a point of failure. Also, it needs a strong communication network, and can have difficulties with scalability and plug-and-play integration when system size increases or in case of adding new DERs (Elmouatamid et al., 2021; Jimeno et al., 2011).

For example, Lin et al. (2015) and Tsikalakis & Hatziargyriou (2011) created centralized energy management strategies maximizing renewable and storage dispatch in microgrids. Likewise, Nemati et al. (2018) utilized genetic algorithms and MILP techniques for centralized EMS.

Table 6.2 Centralized and decentralized control architectures

S. No.	*Control architecture*	*Description*	*Pros*	*Cons*
1.	Centralized	Single controller collects global data and controls all units	Strong global optimization and observability	Vulnerable to failure; not easily scalable
2.	Decentralized	Each unit operates independently using local data	High resilience; minimal communication required	Suboptimal global coordination

- *Decentralized control*

Decentralized control allocates a local controller to every DER or subsystem, allowing independent decision-making according to purely local measurement. Decentralized control eliminates the need to rely on international communication, increases resilience, and eliminates central failure dependencies. Yang et al. (2021) and Wang et al. (2022) commissioned blockchain-based and secure architectures of decentralization in DER-rich networks. Local decision-making can lead to suboptimal system-level behavior, while coordination and synchronization are issues across the network (Elmouatamid et al., 2021).

Reliability studies like those by Garip et al. (2023) indicate that single points of failure may result in the overall failure of the system in case of a fault, including centralized controllers, while decentralized architecture significantly improves fault tolerance and reliability.

Table 6.2 lists centralized and decentralized control architectures with their pros and cons.

6.3.2 Hierarchical and distributed smart grid control architectures

- *Hierarchical control architectures*

Hierarchical control architectures are typically used by microgrids and smart grid networks. These are architected into three separate levels: primary control level, secondary control level, and tertiary control level. Figure 6.3 shows three levels of hierarchical control architectures.

Primary control acts at the shortest timescale to provide local frequency and voltage stability, normally utilizing droop control to facilitate plug-and-play integration of DERs (IJERT, 2020; Rijal et al., 2020).

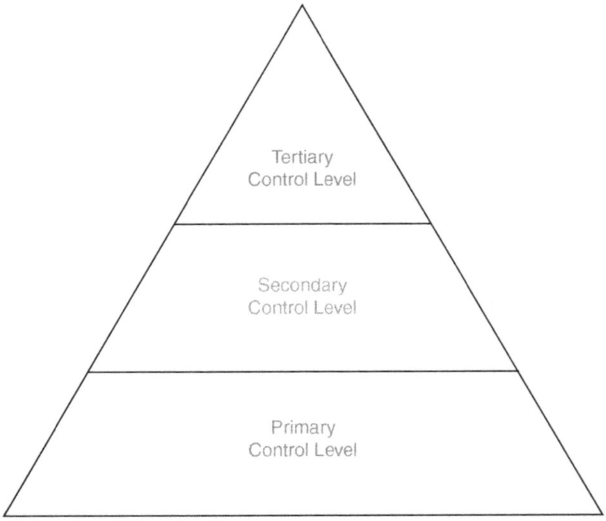

Figure 6.3 Hierarchical control architecture.

Secondary control, on a slower timescale, returns voltage and frequency offsets, synchronizes the microgrid with the main grid when necessary, and optimizes performance throughout the network (IJERT, 2020; Shrivastava & Subudhi, 2020).

Tertiary control is the longest timescale, providing economic dispatch, energy management, and power flow optimization between microgrid and off-site systems (IJERT, 2020; Schulte et al., 2024).

Hierarchical architecture combines local rapid response with coordinating optimization for stability and economic operation.

- *Distributed control structures*

Distributed control designs use local controllers that share data with neighboring nodes to achieve system-wide goals. Distributed control structures provide several advantages:

- Failure resistance, since there is no central point of control.
- Less communication required and more scalability.
- Greater plug-and-play flexibility for DERs integration (Shrivastava & Subudhi, 2020; Han et al., 2021).

As per the research by Bernstein et al. (2015), the COMMELEC architecture uses grid agents (GAs) and resource agents (RAs) that allow scalable control using real-time optimization and abstraction.

Table 6.3 Smart grid control architectures

S. No.	*Framework type*	*Structure and function*	*Key advantages*
1.	Hierarchical Control	Primary → Secondary → Tertiary levels, each with specific timescale and task	Stability, structured optimization, clear layering
2.	Distributed Control	Local controllers coordinating via peer communication	Resilience, scalability, low central dependency
3.	Distributed-Hierarchical (Hybrid)	Hierarchical control across layers, with distributed decision at one or more levels	Best of both: coordinated control, robustness, flexibility

- *Combined distributed-hierarchical control*

Certain frameworks require the combination of hierarchical structure and distributed decision-making to bring the best of both traditions together.

For example, microgrid can be implemented as a distributed consensus-based secondary level, which allows local agents to achieve global goals (Mahmoud et al., 2020).

In direct current (DC) microgrids, hierarchical control can be deployed at primary, secondary, and tertiary levels, whereas higher levels can work in centralized, decentralized, or distributed modes based on communication topology (Shahid et al., 2018).

In networked microgrids (NMGs), hierarchical and distributed architecture is integrated to enable optimal power sharing, coordination of DERs, and market operation in connected systems (Han et al., 2021).

Table 6.3 shows hierarchical, distributed, and hybrid control architectures with their functions and advantages.

6.3.3 Management systems in smart grids

- *SCADA*

SCADA systems form the backbone of remote monitoring, control, and data acquisition over the power grid. They gather real-time information from field devices such as remote terminal units (RTUs) and IEDs and relay it back to central control stations. Operators utilize this information for visualization, control, event logging, and alarm management (Chaudhary, 2025).

SCADA has the core integration function in DMS, facilitating distribution automation (DA) through telemetry, alarm management, and telemetry-based distant control of feeders and substations (EETP, 2025). It is also important for fault detection and isolation and facilitating grid data integration from geographic information system (GIS), meter data management system (MDMS), and other systems for outage and load management (IJERT, 2020).

- *EMS*

EMS is the operational intelligence for generation and transmission systems. EMS processes the information collected by SCADA in a manner that supports real-time decision-making for resource optimization, economic dispatching, and load balancing. EMS makes power system management effective by controlling and monitoring load, resource optimization, system reliability, and automatic control decisions, according to instrumentation authorities (Chaudhary, 2025).

In addition, modern EMS uses more and more advanced machine learning (ML) and deep learning techniques to address the complexities of harmonizing renewables, storage, and Internet of Things (IoT) devices to enhance prediction accuracy and operational effectiveness in smart grids (ResearchGate, 2025).

- *DMS*

DMS is concerned with the lower-voltage distribution network, handling feeders, transformers, and outages at maximum efficiency. It generally uses SCADA for real-time grid information and then automatically sends control commands (EETP, 2025).

Industry leader Hitachi offers utility-focused research emphasizing how DMS improves distribution efficiency and reliability with increasing load uncertainty and DER penetration. Business continuity plan (BCP) operations are enabled through DMS, fault isolation is provided, and supply reliability is maintained through data-driven decision-making (Horii et al., 2017).

Table 6.4 lists different management systems in smart grids with their roles and functions.

Table 6.4 Management systems in smart grids

S. No.	*System*	*Role & functions*
1.	SCADA	Remote real-time data acquisition and control; supports DMS and EMS operations; enables fault detection, telemetry, alarm logging, operator visualization
2.	EMS	Central system for generation/transmission grid operations; optimizes dispatch, load, and generation; integrates advanced analytics and ML tools
3.	DMS	Manages the distribution network; isolates faults and restores service; enhances operational efficiency and resilience; supports data-informed resource upgrades

6.4 AUTOMATION TOOLS AND TECHNOLOGIES

There are different tools and technologies that are being used for automation in smart grids. These tools and technologies are helpful in optimizing the performance of the grid. Automation tools and technologies are very important in smart grids for improving the reliability, efficiency, and robustness of the grid. Figure 6.4 shows different tools and technologies.

Automation tools and technologies are explained below:

6.4.1 Intelligent electronic devices

Intelligent electronic devices are very helpful in automation and control of smart grids. Gungor et al. (2013) observed that IEDs are the key components that support automation and control functions at multiple sites in the power system.

IEDs are advanced microprocessor-based devices. These are used to undertake protective relaying, control, monitoring, and data acquisition functions.

They are able to deal with real-time sensor information and other field equipment. They enable immediate decision-making for fault detection, fault isolation, and system restoration as per the research by Kumar & Singh (2019a, b).

IEDs are capable of interacting with other devices and control systems through standardized protocols like IEC 61850. Zhang & Li (2017) states that they support interoperability and smooth integration within the smart grid system.

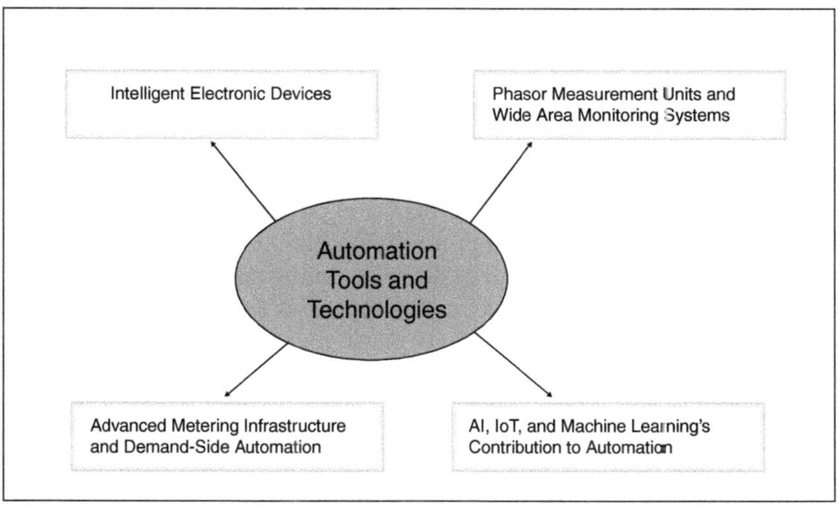

Figure 6.4 Automation tools and technologies.

IEDs also offer remote monitoring and control. This is critical for effectively managing complex and geographically dispersed grid assets (Mishra & Sinha, 2016).

Overall, IEDs are very helpful in automation and control of smart grids.

6.4.2 PMUs and WAMS

PMUs are advanced devices used in modern power systems. They are used for the measurement of electrical waves on an electrical network in real time. PMUs accurately and synchronously measure electrical quantities. These electrical quantities include voltage and current phasors, frequency, and rate of change of frequency, normally from GPS signals (Pham et al., 2014). This ensures synchronization during the collection of time-stamped data from multiple locations. This data can then be analyzed accurately and coherently for grid conditions.

WAMS utilize the capability of PMUs to optimize the performance. They provide an integrated, real-time image of the general health of the power system over extensive geographic spaces. WAMS integrate data from multiple PMUs installed at strategically located positions in the grid. They provide wide-area situational awareness as per the research by Zhao & Wang (2018). They allow operators to examine dynamic performance, detect disturbances, and tune response to contingencies more effectively.

The deployment of PMUs and WAMS strongly enhances the stability, security, and reliability of the power grid. They are helpful in early detection of anomalies such as oscillations, faults, or system instability (Amin et al., 2015).

6.4.3 AMI and demand-side automation

AMI is a complete system that provides two-way communication among utilities and consumers. It uses smart meters and underlying communications networks for its operation. Smart meters capture detailed usage information in real-time. They provide more accurate billing, detecting outages, and improving demand management (Umar et al., 2016). AMI combines hardware, software, and communication technologies. They make dynamic pricing, remote meter reading, and real-time monitoring possible.

Demand-side automation is defined as the use of smart appliances to optimize and control end-users' electricity consumption. Demand-side automation enables customers to respond to price signals or grid conditions by altering usage levels, which contributes to lowering peak demand as well as grid stability (Zhao & Wang, 2017). Automated demand response (ADR) systems are incorporated with AMI to enhance the performance. They have the ability to shed or shift loads automatically during moments of stress, thereby resulting in greater energy efficiency and saving.

Together, AMI and demand-side automation help in energy management. They allow utilities and customers to participate actively in energy management. They increase the flexibility and responsiveness of power systems. They also facilitate the integration of renewable resources and help dynamically balance supply and demand. They serve as the foundation on which efficient, reliable, and sustainable smart grids are built.

6.4.4 AI, IoT, and ML's contribution to automation

The intersection of AI, the IoT, and ML has revolutionized automation across most of the sectors. These sectors include energy, manufacturing, healthcare, and transport. These technologies enable systems to communicate more intelligently and efficiently.

AI allows systems to read complex data and maximize processes autonomously. AI algorithms such as expert systems and reasoning engines facilitate the automation process. They enable predictive maintenance, fault detection, and autonomous control (Russell & Norvig, 2020).

IoT provides interconnection between a large network of sensors, devices, and control systems. They form a real-time data collection network. IoT devices generate vast amounts of data. This data provides information on system performance, environmental conditions, and user behavior. This information is of utmost importance for intelligent automation (Atzori et al., 2010).

ML enables systems to make decisions through patterns. ML algorithms are employed in predictive analytics, anomaly detection, and adaptive control approaches, which lead to more proactive and resilient automation solutions as per the research by Chui et al. (2018).

These technologies present adaptive, predictive, and autonomous systems. These systems can optimize processes, reduce human workload, and improve safety and efficiency.

For example, in smart grids, AI and ML processes IoT sensor data to forecast load demands. They identify potential faults early and control actions automatically according to the research by Li et al. (2019). Similarly, in manufacturing, these technologies present Industry 4.0 automation. It enables flexible production lines and self-optimizing factories.

6.5 APPLICATIONS OF AUTOMATION IN SMART GRIDS

6.5.1 Grid self-healing and smart grids fault management

Self-healing is a core application of smart grid automation. It allows the grid to detect faults automatically, isolate them, and recover from faults. It does all without human intervention to ensure reliability and minimize outages. Zerihun et al. (2022) states that this is achieved by having a network of smart sensors, automated controls, and sophisticated algorithms infused throughout the grid.

- *How grid self-healing works*
    - Fault detection: Sensors and monitoring devices regularly scan grid conditions to optimize the performance. The system detects anomalous patterns in data in real-time on the occurrence of the fault (short circuit, line break, or equipment failure).
    - Fault isolation: Automation controls immediately identify the fault location, which helps to isolate the faulty segment. This quick isolation stops the fault from spreading. It also minimizes larger-scale disruptions.
    - Automatic restoration: Upon isolation, the system diverts electricity around the faulted section by means of automatic switches. It helps to restore power to as many consumers as possible in seconds to minutes (Zerihun, 2025).
    - Continuous learning: Information from every incident refines the grid's models, enhancing performance for subsequent incidents (Zerihun et al., 2022).

These features significantly minimize the time and scale of outages relative to traditional grids' manual fault management. For instance, fault location, isolation, and service restoration (FLISR) technology is found to minimize the number of customers impacted by as much as 45% and outage duration by over 50% (Zerihun, 2025).

6.5.2 ADR in smart grids

ADR is the utilization of advanced communication systems and control technologies. It automatically modifies end-users' electricity usage based on grid conditions such as peak periods or supply shortages. In contrast to the conventional DR, which is time-consuming and relies on human intervention, ADR facilitates real-time distant and automated device and load control to provide faster and more consistent response for sustaining grid stability and efficiency.

- *Important characteristics and benefits of ADR*
    - Peak demand management: Peak demand is managed by ADR via automatic deferral or reduction of electricity use during peaks and thereby shifting the load away from the grid and contributing to the reduction of reliance on peaking power generation plants, which are expensive.
    - Energy efficiency: It maximizes the timing and extent of consumption, avoiding wastage and enhancing overall energy utilization.

- Grid reliability and stability: ADR enhances power system operational stability and resiliency through dynamic balancing of demand and supply.
- Customer convenience: Automation eliminates the need for customers to manually participate but facilitates financial rewards in terms of reduced use during peak periods.

- *Technologies and implementation*

The ADR systems utilize AMI for real-time metering, IoT devices and sensors for smart load control, and communication protocols such as OpenADR for interoperability. Increasingly, AI and ML models are being employed to forecast behavior of demand and optimize response strategies.

- *Case example*

A pilot program in Delhi proved 78% reduction of peak demand from air conditioners by implementing ADR strategies based on LoRa (long-range) communication technology and smart relay devices for appliance control. Economic benefits like delayed infrastructure cost and lower power procurement cost were realized through the program, with payback periods between 1.2 and 3.6 years (Lee, 2025).

6.5.3 Management of renewable energy integration and variability in smart grids

Taking recourse to renewable sources like solar, wind, and hydro is a major challenge as these are intermittent and variable in nature. The variability, i.e., the random behavior in power output fluctuations could undermine grid stability and reliability unless properly addressed.

- *Variability challenges*
    - Supply-side volatility is caused by fluctuating generation due to weather conditions, time of day, and season. For example, solar generation fluctuates with sunshine and cloud cover, while wind generation fluctuates with the variation in wind speeds.
    - Demand-side volatility is about fluctuating consumer electricity consumption. It varies by day and season.

Supply and demand volatility requires advanced management techniques. These techniques help in maintaining a balanced generation and load at all times.

- *Smart grid technologies for variability management*

Smart grids utilize various core technologies for variability management. It includes:

- AMI: It offers real-time consumption information on electricity. It enables improved prediction and response to demand.
- Energy storage systems (ESS): Storage systems such as batteries, pumped hydro, etc., store surplus renewable energy during high generation. This energy is used during low generation times to level out supply fluctuations.
- DR: It encourages customers to change consumption patterns based on grid demands. It enables supply and demand management.
- Advanced forecasting: It utilizes weather forecasts, ML algorithms, and statistical models to predict renewable generation and demand with high accuracy. It enables anticipatory grid management.
- Smart inverters: They regulate DC power from renewable sources to alternate current (AC). They also stabilize the grid by controlling voltage and frequency.
- *Control and optimization techniques*

Smart grids use real-time monitoring and advanced control techniques like:

- Model predictive control (MPC): Makes predictions of future grid states using modeling and optimizes storage and DR control actions.
- Adaptive and hierarchical control: Adapts to different conditions, distributing the control tasks over different levels to supply stability and flexibility.

Utilizing these technologies and techniques together, smart grids create a flexible and secure power system capable of flexing to respond to the variability of renewables, supply a reliable and efficient source of electricity while complementing climate goals.

> "Smart grids manage the uncertainty of renewable energy through forecasting, energy storage, demand response, and adaptive control, creating a strong and efficient power system capable of integrating high penetration of renewables" (Energy Sustainability Directory, 2025).

6.5.4 Microgrid automation and control strategies

Microgrids are localized power systems that can function autonomously or in synchronism with the grid. Automation and control strategies are essential for controlling DERs, maintaining reliability, optimizing energy consumption, and facilitating seamless mode change between grid-connected and islanded modes.

6.5.4.1 *Major control strategies in microgrid automation*

- Hierarchical control: Microgrid control is usually implemented in the structure of hierarchical levels.
  - Primary level control: Provides real-time operation for controlling voltage and frequency with capabilities of droop control for enabling power sharing among DERs.
  - Secondary level control: Recovers from power sharing errors and returns to nominal frequency and voltage after disturbances.
  - Tertiary level control: It maximizes operation, energy management, and economic dispatch, between microgrid and primary grid.

This hierarchical structure enables scalability, modularity, and improved system stability.

- Centralized and decentralized control: Centralized and decentralized controls are given below:
- Centralized control uses a single central controller. It is appropriate for small microgrids but susceptible to single points of failure.
- Decentralized or distributed control allocates control responsibilities across individual DERs or controllers, promoting reliability, flexibility, and scalability.
- Real-time supervisory control: Supervisory control systems continuously monitor the microgrid; controlling transitions between grid-connected and islanded operation modes, optimize energy flow, and power quality assurance.
- Adaptive and intelligent control: Advanced control techniques such as MPC, fuzzy logic, and ML allow for adaptive action against renewable generation and uncertainty in load, forecast errors, and dynamic conditions.
- Protection and fault management: Automation incorporates adaptive protection schemes to disconnect faults rapidly and permit safe running under grid disturbances or islanding mode switching (Zhu et al., 2025).

6.5.5 ELECTRIC VEHICLE CHARGING COORDINATION IN SMART GRIDS

Electric vehicle (EV) charging coordination is important to balance rising electricity demand induced by large-scale EV adoption with grid reliability, efficiency, and stability. Uncoordinated charging of EVs may cause peak loads to rise to high values. It therefore results in the voltage drops and thus losses on the distribution network will increase. Coordinated charging approaches utilize smart grid technology to maximize power consumption, alleviate grid stress, and facilitate integration with renewable energy.

Important concepts of EV charging scheduling:

- Optimal scheduling: Particle swarm optimization (PSO) algorithms are utilized to develop optimized charging schedules. It results in reduced total active power loss and healthy voltage profiles for distribution networks. Coordinated delay or shifting of EV loads to off-peak periods lowers peak demand and increases network efficiency (Chandra, 2024).
- Synchronizing with renewable energy: Smart grids allow the synchronization of EVs' charging with the generation of renewable energy. By coordinating charging times with peak solar or wind power supply hours, EVs become flexible loads that assist in balancing variable renewable output, making the grid more sustainable.
- Real-time control and communication: Smart metering infrastructure (AMI), DMS, and DR systems enable the real-time monitoring and control of EV chargers. Utilities can send price signals or control commands to dynamically optimize charging behavior.
- Grid resilience and load management: Synchronized charging avoids jammed transformers and lines, improves voltage stability, and reduces differences between load and generation. This adds to overall grid resilience, as EV penetration rises (Chandra, 2024).

6.6 COMMUNICATION AND CYBERSECURITY ASPECTS

This section covers the analysis of communication and cybersecurity within smart grid automation systems. It includes:

- Communication protocols (DNP 3, IEC 61850, etc.)
- Cyber-physical vulnerabilities
- Security measures for resilient automation

6.6.1 Communication protocols for automation (DNP 3, IEC 61850, etc.)

Smart grid automation is based on solid communication standards for exchanging real-time data between devices, substations, and control centers.

- DNP 3 (Distributed Network Protocol 3): This standard is popularly deployed in SCADA systems to enable communications between master stations, RTUs, and IEDs. It is reliable in extreme environments. It has been observed that this is not secure by design and requires augmentation such as transport layer security (TLS) or bump-in-the-wire encryption for secure deployment.

- IEC 61850: It is an exhaustive global standard. This is meant for substation automation. Substation elements are represented as objects and provide rapid peer-to-peer communication. They use mechanisms such as GOOSE and sampled values, over high-speed Ethernet for communication.
- Other protocol context: There are other older protocols, such as Modbus and ICCP, which are also utilized in smart grids. These tend to have little resistance to attacks. Higher-level automation systems (such as SCADA) draw significantly on DNP3, IEC 61850, and Modbus according to the research by Sridhar (2012).

6.6.2 Cyber-physical vulnerabilities in automated control systems

Smart grid automation protocols are exposed to cyber-physical threats. These threats can threaten the stability and security of the grid.

- Protocol vulnerabilities: DNP3 and IEC 61850 standards weren't initially designed with robust security. For example, GOOSE messages in standard IEC 61850 can be compromised; specific cases illustrate hackers modifying packets to make circuit breakers trip, leading to cascading outages (WSU Research, 2019).
- Advanced persistent threats (APTs): A two-phase APT attack on an IEC 61850-based substation was illustrated through research utilizing a hardware-in-the-loop testbed. It demonstrates the sophisticated nature of the threat according to the work by Akbarzadeh et al. (2024).
- False data injection (FDI) Attacks: The standard messaging of IEC 61850 enables attackers to inject spurious GOOSE messages. A testbed was created by researchers to emulate such FDI attacks. It proposed a hybrid defense mechanism (sequence content resolver) against them (Hussain et al., 2023).
- Large threat landscape: A 2025 report consolidates IEC 62351 security provisions for IEC 61850 messaging, and catalogues a list of threats (e.g., spoofing, APTs, IoT-based vulnerabilities). It also illustrates overall challenges to grid resilience as per the research by Yaacoub et al. (2025).
- Communication range security gaps: Cyber ranges for smart grids have focused on protocols like DNP3 and Modbus TCP. IEC 61850-based testbeds for cybersecurity are mainly limited to the substation domains (ArXiv, 2025).

6.6.3 Security measures for resilient automation

To address these weaknesses, a range of defense measures has been proposed:

- IEC 62351 Series: These standards provide layered security features. These features include authenticating, encrypting (TLS), authorizing, and securely logging on top of power system communications protocols like IEC 61850 and DNP3 (Wikipedia, 2025).
- Encryption and authentication retrofits: It is essential to add encryption, authentication, and access control to new protocols. Retrofitting these security layers into older systems minimizes their attack surface (Sridhar, 2012).
- Intrusion detection systems (IDS): Distributed IDS with semantic rules can be used to scan DNP3 traffic in real time at substations for detecting anomalies and improving SCADA cybersecurity (Mohan et al., 2024).
- SDN-based IDS frameworks: Software-defined networking (SDN) offers dynamic cybersecurity architectures, such as blocking malicious GOOSE messages by port in an IEC 61850 framework demonstrated through HIL testbeds (Girdhar et al., 2023).
- Periodic vulnerability scans: Periodic proactive measures such as penetration testing, audits, and ongoing monitoring are recommended to identify and repair vulnerabilities in processes, networks, and systems (Gev-Ernova Strategies, 2025).

Table 6.5 presents different risks and security measures.

6.7 CASE STUDIES AND INDUSTRY PRACTICE

6.7.1 DA in urban smart grids

- *What DA does?*

DA is the collection of sensing, communications, control, and analytics capabilities that allow utilities to observe and control the distribution system more akin to the transmission system: automatic fault detection, isolation and service restoration (FLISR), Volt/VAR optimization (VVO), conservation voltage reduction (CVR), feeder reconfiguration, advanced outage management, and DER orchestration through an advanced distribution management system (ADMS). DA integrates field devices (IEDs/feeder controllers), AMI information, RTUs/SCADA interfaces, and an ADMS/DA application layer to automate decisions that are done manually (DOE SGIG lessons; LADWP program).

Table 6.5 Risks and security measures

S. No.	*Area*	*Key risks*	*Security measures*
1.	DNP3/ Modbus/ IEC 61850	Lack of built-in authentication/ encryption; protocol-specific attacks (e.g., FDI, spoofing)	IEC 62351 retrofits; TLS; secure logging; IDS
2.	Cyber-Physical Attacks	APTs, replay/spoofing, false data injection, cascading failures	Security testbeds; APT simulations; ICS-specific mitigation
3.	Detection & Monitoring	Stealthy attacks bypassing static defenses	Distributed IDS; anomaly detection; SDN-enabled defense
4.	Ongoing Management	Emerging threats, human error, policy gaps	Vulnerability audits; training; compliance with IEC 62351 & EC 62443

- *Why DA is important for dense urban grids?*

Urban distribution needs to reconcile extremely high load density, scarce space for new substations, strict outages and emission constraints, and high behind-the-meter DER growth. DA facilitates non-wires alternatives(NWAs), quicker outage restoration, capital postponement, and enhanced integration of storage and DR to control local peaks (Con Edison BQDM; LADWP).

- *Industry representative cases*
  - Enel-Telegestore (Italy): Large-scale AMI deployment (early 2000s) that automated meter reading, supported real-time profiling and fraud avoidance, and paved the way for DA and customer-facing services. Enel's initiative illustrates how an extensive AMI program becomes the foundation for DA and transparency in big cities (Enel, Telegestore).
  - Con Edison-Brooklyn-Queens Demand Management (BQDM): An urban NWA in which Con Edison utilized DERs, DR, and special customer incentives to postpone a scheduled $1.2 billion substation upgrade. BQDM is a classic instance of employing DA, DR aggregation and strategic investments in deferring expensive reinforcement in an urban setting (Con Edison BQDM program materials; ORNL profile).
  - LADWP smart grid regional demonstration program: Demonstrations funded by DOE (campus pilots such as UCLA) that implemented FLISR, VVO, AMI, and ADMS integrations to enhance reliability and pave the way for city-scale DA features (LADWP DOE SGRDP).

- Urban-level strategies (Singapore Grid 2.0/SP Group): Singapore's Grid-2.0 planning emphasizes how compact city-state grids concentrate DA on distribution visibility, real-time control, DER integration, and resilience as an element of a Smart Nation plan (NTU/SP Group materials).

6.7.1.1 *Most important technical lessons from DA pilots in cities*

- AMI + ADMS integration is a building block. Without timely, accurate meter/feeder data, the higher-level automation cannot yield value (Enel; DOE SGIG).
- Non-wires alternatives need strong measurement and verification (M&V) and contractual/market arrangements to compensate third-party providers. BQDM illustrates the policy and program design necessary to recover value.
- Interoperability, standards (IEC 61850/DNP3/CIM for data models), and security need to be planned upfront; retrofits are expensive (DOE SGIG lessons).

6.7.2 Wide-area control and stability improvement in transmission grids

- *What "wide-area" control (WACS/WAMS/WADC) does?*

Wide-area measurement and control systems (WAMS) use time-synchronized PMUs and high-speed communication to deliver system-level state measurements (voltage/angle/frequency) in large geographical areas. Wide-area damping controllers (WADC), under-frequency load-shedding (UFLS) coordination, and automatic emergency control utilize PMU streams to monitor inter-area oscillations, analyze dynamic stability, and initiate coordinated controls (FACTS, SVC/TCSC, generator excitation/PSS, fast energy resources) to damp oscillations or stop cascading events (Phadke & Bi, 2018; PNNL demonstrations).

- *Why it is important?*

With higher geographic interconnection and more renewable penetration (lower inertia), inter-area oscillations and rapid stability events can spread rapidly. Local SCADA by itself does not have the synchronization and refresh rate to see these modes; PMU-based WAMS/WADC enhances detection and facilitates quick coordinated responses to maintain stability (DOE/PNNL work; WECC analyses).

- *Important field tests and research*
  - PNNL/DOE demonstrations of active damping control and wide-area damping: Pacific Northwest National Laboratory and collaborators demonstrated active damping of inter-area modes with PMU data and wide-area controllers, demonstrating quantifiable improvements in dynamic response and oscillation damping in test deployments.
  - PMU deployments and operator tools (RTDMS, Oscillation Detection): PMU-based real-time analysis platforms have been installed to offer oscillation detection, modal analysis, and operator alarms; these have been applied to transmission control centers to enhance situational awareness and response to events (PNNL RTDMS; Phadke & Bi, 2018).
  - WAMS-based UFLS and emergency control optimization: Research and pilots (Haes Alhelou et al., 2019) used WAMS information to create optimal multi-stage under-frequency load shedding schemes more responsive to system conditions than conventional static UFLS blocks.
- *Practical engineering lessons from transmission pilots*
  - Data quality and latency are important: PMU sampling and communications latency impose boundaries on what control loops may safely perform; latency, GPS dependability, and data management were frequent engineering concerns in pilots (DOE/PNNL).
  - Control design needs to be resilient to cyber/communication failure: Adaptive and "dynamic-loop" WADC designs that can reroute measurement/control channels in the event of a suspected attack or communications failure enhance resilience (Patel et al.; adaptive WADC research).
  - Operator integration and governance: Wide-area controls have to be integrated into current EMS/SCADA workflows and operator practices; pilots demonstrated operator trust establishment, obvious human-in-the-loop rules and visualization are crucial prior to enabling automated wide-area actions (PNNL reports; WECC analyses).

6.7.3 Microgrid and community-level automation projects

- *Microgrid automation*

Microgrid controllers (MICs or MGCCs), microgrid energy management systems (microgrid EMS), and hierarchical control (primary/secondary/tertiary) with local DER controllers manage PV, storage, dispatchable generators, and controllable loads to run grid-connected or islanded. Automation includes state estimation, island detection, black-start, economic dispatch, and rapid real-time controls for voltage/frequency (droop, virtual inertia) (UCSD, DOE case studies).

- *Real-world examples*
  - Borrego Springs (California) SDG&E Community Microgrid: Borrego Springs is a rural community where SDG&E deployed a renewable-based storage-enabled community microgrid with control that successfully islanded during repeated tests and real outages (served ~2,800 customers for hours during planned tests and storms). The demonstration of community-scale automation and storage provides resilience and support for distribution operations (SDG&E Borrego Springs Project).
  - Blue Lake Rancheria (California) Tribal Community Microgrid: Blue Lake Rancheria installed solar, battery storage, and operator-less MGMS (Siemens Spectrum Power) to enable islanded operation for essential services (casino, tribal buildings). This DOE-funded project emphasizes community resilience, operator training, and the benefit of microgrids for emergency readiness.
  - University of California San Diego (UCSD) Microgrid/Campus DER Orchestration: Microgrid and institutional research at UCSD shows intricate microgrid economics and control: Campus DER aggregation, ancillary service participation, and need to reconcile "missing-money" economics for microgrid investments with decarbonization objectives (UCSD microgrid studies).
  - Brooklyn Microgrid (LO3 Energy/TransActive Grid): Peer-to-peer trading pilot: A technology-intensive pilot combining local DER aggregation with blockchain-based settlement to test peer-to-peer energy transactions at the neighborhood level. The test drew valuable lessons on market design, regulatory friction, and measurement/settlement accuracy rather than demonstrating blockchain as an essential control mechanism (LO3/Brooklyn Microgrid studies).
- *Microgrid automation lessons from deployments*
  - Control maturity and testing: Microgrids require phased testing (HIL-hardware in the loop, black-start exercises, islanding tests) prior to commercial operation; Borrego and Blue Lake conducted repeated islanding tests that confirmed both control logic and restoration procedures.
  - Commercial and regulatory models: Commercial and regulatory models are essential; UCSD/NREL studies indicate that economics frequently depend on layering services (resilience

value, demand charge reduction, wholesale market participation) to make capital cost recoverable. Pilot outcomes confirmed that without definite revenue streams or policy incentives, most microgrids cannot achieve bankable returns.
- Community engagement and operations: Where end-users (municipal, tribal, or campus) were engaged in planning, projects saw greater success for resilience and social acceptance. Well-defined roles for operations, maintenance, and emergency procedures are required.

6.7.4 Lessons learned from pilots: cross-cutting themes and recommendations

Top recurring lessons from DOE, CEC, NREL, and large pilot:

- Plan standards, interoperability, and data models early: Data needs to be shared by AMI, ADMS, EMS, DMS, and DER controllers using interoperable models (CIM/IEC 61850, where applicable). Retrofitting interoperability is expensive (DOE SGIG; Enel Telegestore lessons).
- Design definite performance metrics and M&V upfront: Pilots that measured demand-reduction, reliability (SAIDI/SAIFI), and customer benefits yielded transferable business cases (Con Edison BQDM; CEC microgrid lessons).
- Cybersecurity and communication resilience: These are not discretionary. It is required to implement secure communications, segmentation, IDS/monitoring, and resiliency measures up front; pilots determined that cyber hardening and secure architecture have to be budgeted and field-tested (DOE & PNNL WAMS guidance; SGIG lessons).
- Human factors and operator integration: Operator training, visualization, and incremental automation establish trust; least-disruptive automation modes (assistive, advisory) enable operators to accept automated control before complete autonomy (DOE SGIG; PNNL).
- Market and regulatory structures: There is a need to facilitate value capture. Non-wires alternatives, DER aggregation, and microgrids need tariff and market structures (rules for procurement, interconnection process, incentives); BQDM and NREL DER aggregation reports identify this as the most prevalent barrier.
- Testing and staged deployment: HIL testing, staged islanding tests, and multi-season tests uncovered failure modes early and accumulated operational knowledge. Long-term observation is necessary to measure actual benefits.

6.8 SMART GRID AUTOMATION CHALLENGES AND FUTURE RESEARCH DIRECTIONS

Although multiple advantages, smart grid automation faces many challenges. These challenges include scalability and interoperability challenges, integration of autonomous control with AI, issues of edge and cloud-based control paradigms and challenges arise due to vision towards fully autonomous smart grids. Table 6.6 lists different challenges and future research directions.

6.9 CONCLUSION

6.9.1 Summary of key findings

This book chapter emphasizes that the development of automation and control in smart grids is crucial in realizing efficient, reliable, and sustainable energy systems. The challenges of scalability and interoperability rise due to the heterogeneity of devices and communication protocols need to be addressed by the creation of universal standards and flexible architectures. Although multiple advantages of AI, such as autonomous decision-making, predictive analytics, and system optimization, there are concerns in terms of model reliability, explainability, and cybersecurity.

The edge and cloud paradigms of computing increase the responsiveness and scalability of smart grid control systems but introduce security and latency issues. So there is a need for innovative hybrid architectures and secure communication protocols.

Eventually, the completion of fully autonomous, self-healing, adaptive smart grids will continue to require research breakthroughs in AI resilience, cybersecurity, and regulatory approaches.

6.9.2 Future outlook for automation and control in next-generation smart grids

The next generation of smart grid automation will be obtained by greater incorporation of sophisticated AI algorithms, distributed control frameworks, and secure, scalable communication systems. Self-learning, adaptive design of control systems will allow grids to manage variable loads. The variability of renewable generation can also be managed in an adaptive manner, increasing resilience and efficiency.

Federated learning and other privacy-preserving methods will allow secure, distributed data analysis. They focus on building trust and enabling broader deployment. The transition to complete autonomous smart grids will rely on overcoming existing problems of standardization, cybersecurity, and regulatory acceptability.

Table 6.6 Challenges and future research directions

S. No.	*Section*	*Challenges*	*Future research directions*
1.	Scalability and Interoperability	Exponential growth in devices and data streams (sensors, smart meters, DERs). Complexity in communication and data management with heterogeneous systems. Integration issues due to lack of common standards and legacy systems. Risk of performance decline as network scales	Development of adaptive and scalable data frameworks for storage, processing, and analysis. Creation of universal communication protocols and widespread adoption of standards like IEEE 2030.5. Use of middleware solutions to integrate legacy and modern systems. Leveraging edge computing to reduce network load and latency. Designing dynamic, self-organizing network architectures
2.	Autonomous Control Integration with AI	Ensuring AI model reliability, safety, and predictability under uncertain and extreme conditions. Lack of transparency in AI-driven decisions affecting grid safety. Risk of unstable autonomous actions without proper fail-safes	Development of Explainable AI (XAI) for operator trust and interpretability. Research into resilient, adaptive AI algorithms for real-time decision-making. Integration of fail-safe mechanisms and robust testing frameworks. Creation of trustworthy AI governance models for critical infrastructure
3.	Edge and Cloud-Based Control Paradigms	Latency issues in real-time operations. Synchronization and consistency challenges between edge and cloud. Expanded cybersecurity risks due to distributed architecture. Bandwidth management in large-scale deployments	Design of hybrid edge-cloud architectures for balanced performance. Development of secure and low-latency communication protocols. Research into federated learning for decentralized, privacy-preserving model training. Advanced cyber defense strategies for distributed systems
4.	Vision Towards Fully Autonomous Smart Grids	High complexity in enabling self-optimizing, self-healing, and self-protecting operations. Cybersecurity vulnerabilities of autonomous systems to attacks. Regulatory and policy gaps in standards, liability, and safety. Risk of over-dependence on AI without human oversight	Development of self-tuning, adaptive control algorithms. Embedding AI-driven intrusion detection and defense mechanisms. Establishing comprehensive regulatory and safety frameworks. Building cyber-resilient autonomous infrastructures. Promoting human-in-the-loop oversight for critical operations

Explainable AI and strong safety protocols will be essential to build trust with operators and maintain system stability. The industry players, policymakers, and researchers will need to come together to design standards, devise policies, and safeguard infrastructure.

In total, future smart grids will increasingly rely on interoperable, intelligent, and secure control systems. These control systems will be fueled by leading-edge AI, edge-cloud convergence, and open standards of interoperability. They will ultimately support a more resilient and sustainable energy future with global clean energy transition and grid modernization goals.

REFERENCES

Abdullah, A. A. (2022). Smart Grid (SG) Properties and Challenges: An Overview. *Energy and Built Environment*. 3(3): 285–298.

Akbarzadeh, A., Erdodi, L., Houmb, S. H., & Soltvedt, T. G. (2024). Two-Stage Advanced Persistent Threat (APT) Attack on an IEC 61850 Power Grid Substation. *International Journal of Information Security*. 23:2739–2758.

Amin, S., et al. (2015). Wide Area Monitoring Systems (WAMS): Technologies and Applications. *IEEE Transactions on Smart Grid*, 6(1), 250–259.

Atzori, L., Iera, A., & Morabito, G. (2010). The Internet of Things: A Survey. *Computer Networks*, 54(15), 2787–2805.

Bernstein, A., Reyes-Chamorro, L., Le Boudec, J. Y., & Paolone, M. (2015). A Composable Method for Real-Time Control of Active Distribution Networks with Explicit Power Setpoints (Part I): Framework. *Electric Power Systems Research*, 125, 254–264.. 10.1016/j.epsr.2015.03.023

California Energy Commission. (2024). *Lessons Learned from Energy Commission Microgrid Projects: Best Practices from EPIC PON-14-301 Grant Recipients.* CEC-500-2024-067.

Chandra, I. (2024). Optimal Charging Schedule Coordination of Electric Vehicles in Smart Distribution Grid. *Indonesian Journal of Electrical Engineering and Computer Science*, 11(1), 82–89.

Chaudhary, N. (2025). *Electrical Energy Management System and SCADA in Power System.* Instrumentation Tools.

Chen, L., & Wang, Y. (2020). Advances in Automation and Control for Smart Grids. *IEEE Transactions on Power Systems*, 35(2), 1234–1245.

Chui, M., Manyika, J., & Miremadi, M. (2018). AI, Automation, and the Future of Work. *McKinsey Quarterly*.

Electrical Engineering Portal (EETP). (2025). *SCADA as the Heart of a Distribution Management System.* EEP Technical Articles.

Elmouatamid, A., Ouladsine, R., Bakhouya, M., El Kamoun, N., Khaidar, M., & Zine-Dine, K. (2021). Review of Control and Energy Management Approaches in Micro-Grid Systems. *Energies*, 14(1), Art. no. 168. doi: 10.3390/en14010168.

Enel. (2005). The Telegestore Project in ENEL's Metering System. *ResearchGate / Enel Technical Report.*

Energy Sustainability Directory. (2025). *How Do Smart Grids Handle Variability?* [Online]. Available at: https://energysustainabilitydirectory.com/how-do-smart-grids-handle-variability/

Garip, S., Bilgen, M., Altin, N., Özdemir, S., & Sefa, İ. (2023). Reliability Analysis of Microgrids: Evaluation of Centralized and Decentralized Control Approaches. *Electric Power Components and Systems*, 51(19), 2319–2338.

Gev-Ernova Strategies. (2025). Securing Smart Grids: Strategies & Best Practices. *Gev-Ernova Energy Security Report.*

Girdhar, M., Hong, J., Su, W., Herath, A., & Liu, C. C. (2023). SDN-Based Dynamic Cybersecurity Framework of IEC-61850 Communications in Smart Grid. *Preprint published on arXiv.*

Gungor, V. C., et al. (2013). Smart Grid Technologies: Communication Technologies and Standards. *IEEE Transactions on Industrial Informatics*, 9(1), 528–540.

Haes Alhelou, H., Hamedani-Golshan, M. E., Njenda, T. C., & Siano, P. (2019). Wide-Area Measurement System-Based Optimal Multi-Stage Under-Frequency Load-Shedding. *Applied Sciences*, 9, 508.

Han, Y., Ning, X., Yang, P., & Guerrero, J. M. (2021).Distributed Cooperative Control and Energy Management Of Networked Microgrids: A Review. *IET Renewable Power Generation*, 15(16), 3588–3607.

Horii, H., Iimura, M., Kirihara, K., Hayasaka, Y., & Yamabe, K. (2017). Work on Supervisory Control and Monitoring Systems in the Transmission and Distribution Sectors. *Hitachi Review*, 66(7).

Hossain, E., & Muhammad, G. (2018). The Role of Automation and Control in the Smart Grid Vision. *IEEE Communications Surveys & Tutorials*, 20(4), 3055–3074.

Huang, G., Wang, J., Chen, C., Guo, C., & Zhu, B. (2017). System Resilience Enhancement: Smart Grid and beyond. *Frontiers of Engineering Management*, 4(3), 271–282.

Hussain, S. (2023). A Novel Hybrid Methodology to Secure GOOSE Messages in IEC 61850. *Scientific Reports*. 13, 27157.

IJERT. (2020). The Role of Advanced Distribution Automation in Smart Grid. *International Journal of Engineering Research & Technology*. 9(2), 148–151. doi: 10.17577/IJERTV9IS020055.

Jimeno, J., Anduaga, J., Oyarzabal, J., & Gil de Muro, A. (2011). Architecture of a Microgrid Energy Management System. *European Transactions on Electrical Power*, 21(2),1142–1158. doi: 10.1002/etep.443.

Khan, S., Malik, M., & Ahmed, R. (2018). Evolution of Automation in Power Systems. *International Journal of Electrical Power & Energy Systems*, 99, 123–130.

Kumar, A., & Singh, R. (2019a). Automation Technologies and Tools in Smart Grids: Intelligent Electronic Devices (IEDs). *Journal of Power Technologies*, 99, 45–54.

Kumar, A., & Singh, R. (2019b). Objectives and Scope of the Chapter on Smart Grid Automation and Control. *Journal of Power Technologies*, 99, 1–10.

Lee, S. (2025). *Automated Demand Response in Smart Grids*. Number Analytics.

Li, F., et al. (2019). AI and IoT for Smart Grid Management: A Review. *IEEE Transactions on Smart Grid*. 10(4), 4034–4045.

Li, Y., & Wang, J. (2020). Automation and Control Technologies for Smart Grids: Enabling a Sustainable Future. *Renewable and Sustainable Energy Reviews*, 128, 109895.

Lin, Y., Wang, J., & Liu, C. (2015). Centralized Energy Management Strategy for Microgrids with Renewable Energy Sources and Energy Storage Systems. *IEEE Transactions on Smart Grid*, 6(3), 1234–1243.

Mahmoud, M. S., Rasheed, M. A. U., & Guerrero, J. M. (2020). Consensus-Based Distributed Control and Energy Management of Networked Microgrids: A Review. *Energies*, 13(7), Art. 1668.

Mansouri, S., Badr, M. A., Al-Farsi, A., Numair, A. K. A., et al. (2025). An In-Depth Survey of Latest Progress in Smart Grids: Paving the Way for a Sustainable Future through Renewable Energy Resources. *Journal of Electrical Systems and Information Technology,* 12, Article 9.

Mishra, S., & Sinha, S. (2016). Role of Intelligent Electronic Devices in Smart Grid Automation. *International Journal of Electrical Power & Energy Systems*, 78, 451–460.

Mohan, S. N., Ravikumar, G., & Govindarasu, M. (2024). Distributed Intrusion Detection System using Semantic-based Rules for SCADA in Smart Grid. *Preprint published on arXiv.*

Nemati, M., Braun, A., & Tenbohlen, S. (2018). Optimal Energy Management in Microgrids Using Genetic Algorithms and Mixed-Integer Linear Programming. *Applied Energy*, 210, 944–963.

Patel, S., Govindarasu, M., & Wang, C. (2019). Cyber-Physical Modeling and Resilient Control for Smart Grid Communications. *IEEE Transactions on Smart Grid*, 10(5), 5238–5247.

Phadke, A. G., & Bi, T. (2018). Phasor Measurement Units, WAMS and Their Applications in Protection and Control of Power Systems. (Technical review / paper).

Pham, H., et al. (2014). Phasor Measurement Units and Their Role in Power System Monitoring. *Electric Power Systems Research*, 113, 100–108.

ResearchGate. (2025). *Energy Management Systems (EMS) in Smart Grid and ML/DL Integration*. ResearchGate Publication.

Rijal, B., Bomzon, K., Dahal, R., & Adhikari, S. (2020). A Review on Hierarchical Control of AC Microgrids. *International Conference on Role of Energy for Sustainable Social Development in 'New Normal' Era*, December 28–29, Kathmandu, Nepal.

Russell, S., & Norvig, P. (2020). *Artificial Intelligence: A Modern Approach*. (4th ed.). Pearson.

San Diego Gas & Electric (SDG&E). (2020). *Borrego Springs Microgrid- Project Report and Islanding Test Summary*, SDG&E, 2017–2021.

Schulte, F., Blank, J. H., & Sonnenschein, M. (2024). Microgrid Energy Management Systems: A Critical Review on Methods, Solutions, and Prospects. *Energy Informatics*, 7(1), Art. 15.

Shahid, M. U., Khan, M. M., Xu, J., Hashmi, K., Habib, S., Mumtaz, M. A., & Tang, H. (2018). A Hierarchical Control Methodology for Renewable DC Microgrids Supporting a Variable Communication Network Health. *Electronics*, 7(12), 418.

Shrivastava, H. & Subudhi, B. (2020). Comprehensive Review on Hierarchical Control of Cyber-Physical Microgrid System. *IET Generation, Transmission & Distribution,* 14(24), 6397–6416. doi: 10.1049/iet-gtd.2020.0971.

Zerihun, T. A. (2025). One Step Closer to a Self-Healing Grid. SINTEF Blog, Oct. 27, 2022, updated Apr. 4, 2025. [Online]. Available at: https://blog.sintef.com/energy/one-step-closer-to-a-self-healing-grid/?utm_source=chatgpt.com

Song, E. Y., FitzPatrick, G. J., Lee, K. B. (2017). Smart Sensors and Standard-Based Interoperability in Smart Grids. *IEEE Sensors Journal,* 17(23), 7622–7634.

Sridhar, S. (2012). Cyber-Physical System Security for the Electric Power Grid. *PowerCyber Technical Report.*

Tsikalakis, N. G., & Hatziargyriou, N. D. (2011). Centralized Control for Optimizing Microgrid Operation. *IEEE Transactions on Energy Conversion,* 26(2), 419–427.

Umar, M., et al. (2016). Advanced Metering Infrastructure and Its Role in Smart Grid. *International Journal of Electrical Power & Energy Systems,* 78, 1–10.

Wang, H., Chen, L., & Li, Y. (2022). Artificial Intelligence-Based Energy Management Framework for Renewable-Rich Microgrids. *IEEE Access,* 10, 45678–45691.

WSU Research. (ca. 2019). Vulnerabilities in IEC 61850 GOOSE Protocol: GOOSE Packet Manipulation. *Electrical Engineering Report.*

Yaacoub, J. P. A., et al. (2025). Toward Secure Smart Grid Systems: Risks, Threats, and Automated Security Measures. *Future Internet,* 17(7), 318.

Yang, X., Zhang, Y., & Wang, J. (2021). Data-Driven Energy Management and Optimization for Smart Microgrids. *Energy,* 228, Art. no. 120597.

Zerihun, T., Farahmand, H., & Liu, J. (2022). Two Novel Current-Based Methods for Locating Earth Faults in Unearthed Ring Operating MV Networks. *Electric Power Systems Research,* 213, 108698.

Zhang, L., Sun, Y., & Chen, M. (2019). Control Strategies for Smart Grid Automation. *IEEE Transactions on Smart Grid,* 10(2), 1972–1981.

Zhang, Y., & Li, J. (2017). IEC 61850 Based Communication for Intelligent Electronic Devices in Smart Grids. *IEEE Transactions on Smart Grid,* 8(4), 1753–1762.

Zhao, C., & Wang, Y. (2017). Demand Response and Smart Metering: Technologies and Applications. *IEEE Transactions on Power Systems,* 32(4), 2871–2879.

Zhao, C., & Wang, Y. (2018). Implementation of Wide Area Monitoring System with PMUs for Power System Stability. *IEEE Transactions on Power Delivery,* 33(4), 1658–1667.

Zhao, H., & Li, X. (2019). Role of SCADA in Modern Power System Automation. *Electric Power Systems Research,* 170, 177–185.

Zhu, X., Li, Y., & Wang, J. (2025). Real-time Supervisory Control Strategy for a Grid-Connected Microgrid Integrating Renewable Energy. *Renewable Energy,* 218, 345–359.

Chapter 7

Cybersecurity and data privacy in smart grids

Karun Handa, Manish Kumar Sharma, and Ramandeep Sandhu

7.1 INTRODUCTION

Electric energy is viewed as the backbone of contemporary society. Nowadays, almost everything is powered by electricity, such as household appliances, industries, and even our transportation systems. The traditional power grid that was developed several decades ago was primarily concerned with the one-directional supply of electricity to the end consumers through the generation plants. It was very unintelligent and inflexible. As the population grows, cities become more popular, and the need to develop renewable energy rises, the conventional grid has begun to be placed under severe constraints. This has prompted the development of the smart grid. A smart grid is fundamentally a combination of the traditional power system and digital technologies like information and communication technology (ICT), Internet of Things (IoT) devices, and artificial intelligence (AI). In simple words, it is a grid that has been modernized so that it is capable of sensing, analyzing, and reacting to changing energy conditions in a real-time manner [1].

The smart grid is highly appealing in its promise to help save energy, cut down on waste, and make the most of renewable energy. For example, in the event of an unexpected disconnection of wind power, the smart grid is capable of ensuring the supply at that particular moment by balancing other energy sources, or when people are parking their electric vehicles at night, the grid could automatically control the load balance. Smart meters at home, where the utility companies have much more detailed usage information, so they can design more efficient tariff regimes, and it also gives the consumer a little control over how much electricity is used. In a nutshell, the smart grid is not only the electricity distribution, but the creation of an interactive and high-technology energy system [2].

This intelligence comes at a cost. The heavy reliance on data and connectivity creates other risks that were never captured in traditional grids. All communication links, sensors, and smart meters can be a source of a threat. In the absence of strong security controls, malicious agents are likely to steal

 DOI: 10.1201/9781003689430-7

or manipulate sensitive data through potential entry points. In one instance, by tracking smart meter data, an attacker can be aware of the unoccupancy of a residence, thereby posing instant security threats. Similarly, when the attackers introduce false information into the network, they have the ability to disrupt load balancing and even cause large-scale power outages. Various sources have pointed out those problems in detail, especially attacks based on false data injection (FDI) and authentication attacks that may make the grid operators confused [3, 4]. This demonstrates that cybersecurity and data privacy are not minor aspects but are the pillars of any smart grid implementation.

The issue is that a smart grid is much more difficult to secure than a traditional IT system. In an ordinary computer network, when a site or a server is offline, the most probable consequence is an inconvenience or data loss. The repercussions are far more extreme in a smart grid. A successful cyberattack may cause direct interruption to the electrical power of industries, airports, and hospitals. The most popular and described case of a widespread incident is the 2015 Ukraine blackout in which hackers targeted substations remotely, shutting hundreds of thousands of people off. In more recent times, the ransomware attacks on power utilities in the United States and Europe have demonstrated that attackers are not only able to, but also intent on exploiting such systems. Research surveys published in 2023 and 2024 also assert that centralized architectures render the grid susceptible to denial-of-service attacks and ransomware [5]. Such accidents and research are important alerts that cybersecurity in smart grids is not to be overlooked [6].

Another serious issue of equal importance is privacy. The highly specific information collected by smart meters can reveal personal habits and routines. It has been found in studies that the application of data mining techniques on the data of electricity usage allows one to identify individual appliances being used and the time of use. The data is highly susceptible to misuse. Insurance companies might try to use it for profiling customers, or criminals may exploit it for planning burglaries. Recent literature has called this the privacy paradox of blockchain: although blockchain provides transparency and integrity, it can still leak metadata and allow profiling of users [7, 8]. Therefore, building trust in the smart grid requires guaranteeing that user information is kept confidential and not accessed by unauthorized parties.

The purpose of this chapter is to address the above issues by bringing to light some of the most problematic issues that still persist despite the breakthroughs between the years 2022 and 2025. Many research works have proposed new cryptographic techniques, intrusion detection systems, and even blockchain-based frameworks to enhance security in federated learning [9, 10]. However, practical deployment of these solutions still faces challenges. Issues such as scalability, resource limitations in edge devices, lack of standardized regulation, and privacy issues associated with existing blockchain architectures are open issues [11, 12]. Highlighting these issues, the chapter attempts to give a general impression of the scenario in the domain and the direction it has to take. To be able to have an effective and meaningful discussion, this chapter shall address five key areas of issue as shown in Figure 7.1.

- **Integrity and authentication failures:** How FDI and lack of strong authentication undermine the trust in grid communications.
- **Centralized architecture vulnerabilities:** Why relying on central servers creates a single point of failure, and how decentralization is becoming necessary.
- **Privacy limitations in blockchain:** The paradox of blockchain transparency and how it may still leak user behavior data.
- **Resource constraints in edge devices:** The challenge of applying strong cryptography on low-power devices like smart meters.
- **Regulatory and interoperability gaps:** The lack of unified standards and policies creates uneven levels of protection across the grid.

The following sections will perform a detailed exploration of each of these issues, including recent academic studies, suggested solutions, and ongoing research challenges. The intent is not to describe the smart grid as fundamentally insecure, but rather to recognize that not solving these security and privacy issues will keep the smart grid from achieving its full potential. The main intent of this chapter is to stress that a secure and privacy-aware smart grid is the foundation of a sustainable, reliable, and trustworthy future energy infrastructure.

7.2 SMART GRID CYBERSECURITY LANDSCAPE

The concept of the smart grid is often explained as a modern version of the traditional power grid. In simple words, it is an energy system that not only transmits electricity but also communicates. The communication is between power plants, substations, smart meters, sensors, and even consumers. This two-way communication is what makes the grid more intelligent compared to the old conventional system. However, at the same time, this very communication also opens the door for cyber threats. Before going into details of the security issues, it is important to first look at how the smart grid architecture is structured. A typical smart grid has three main parts as shown in Figure 7.2. The first part is generation, where energy is produced in power plants. This can include coal, gas, hydro, nuclear, and more recently renewable sources like solar and wind. The second part is transmission and distribution, where high voltage lines carry the power from generation units to

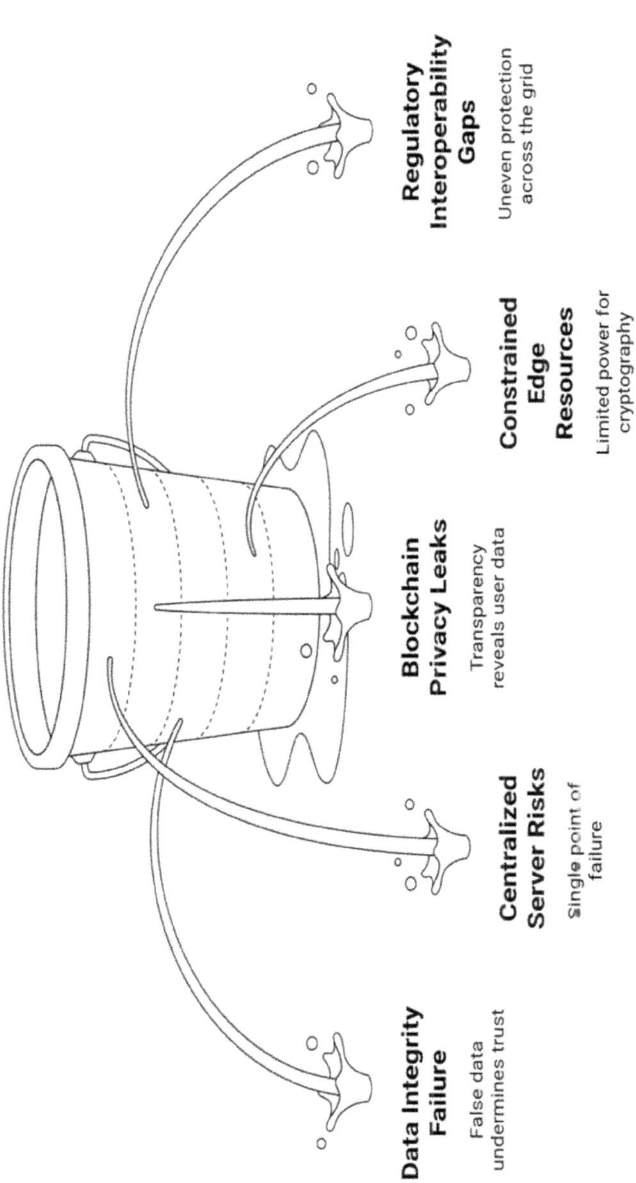

Figure 7.1 Key issues in smart grids.

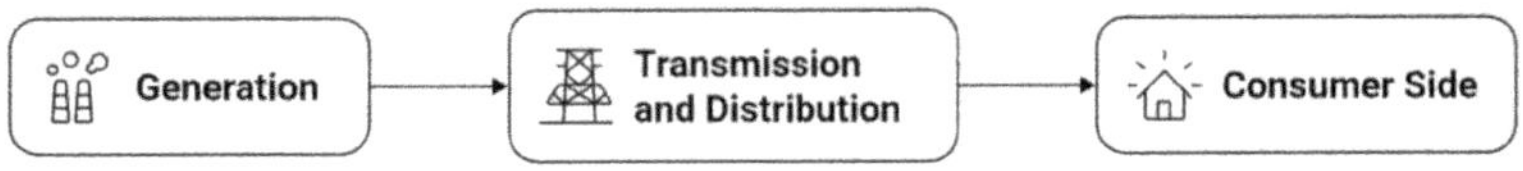

Figure 7.2 Smart grid generation.

cities and towns, and later distribute it to homes and industries. The third part is the consumer side, where electricity is actually used.

What makes the system "smart" is the introduction of ICT and IoT devices in all three parts. Generation units are now embedded with sensors that monitor performance in real time. Advanced control devices are used to generate responses in case of faults in the transmission system. On the consumer side, smart meters record and send detailed usage data back to utilities. Even appliances in homes are slowly becoming part of this connected environment. This tight integration of power and information systems has many benefits, but also creates a large attack surface. Communication channels that were earlier closed are now open to internet-based control. Smart meters, which are installed in millions of homes, can be targeted to manipulate billing data or to disconnect power illegally. Substations, which are the backbone of transmission, can be attacked to disturb frequency or overload transformers. Data repositories that store consumption information can be leaked, resulting in privacy loss of consumers. Control centers, which act as the brain of the grid, are especially sensitive because if they are compromised, the entire grid can be affected. Researchers have noted that almost every layer of the smart grid is now exposed to cyber vulnerabilities. Different types of attacks have been observed and studied in smart grid environments. One of the common attacks is denial of service (DoS). In this attack, the communication channels are flooded with unnecessary traffic so that the genuine control messages are delayed or blocked. In a smart grid, even a small delay can create big problems because the grid operates in real time. Another serious attack is FDI. In this case, the attacker sends manipulated measurement data to the control center. If the control center believes this false data, it can take wrong decisions, which can result in wrong load distribution or even blackouts. Malware-based attacks are also a concern. The well-known Stuxnet incident has already shown how malware can enter industrial control systems. Smart grids, which rely on similar supervisory control and data acquisition (SCADA) systems, can face similar risks. Insider attacks are equally dangerous. Employees with authorized access can misuse their privileges to leak or alter sensitive information. These different threats can be grouped into a taxonomy. Broadly, they can be divided into four categories listed in Figure 7.3. The first category is availability attacks, where the main goal is to disrupt the availability of services, such as DoS. The second is integrity attacks, where data is changed

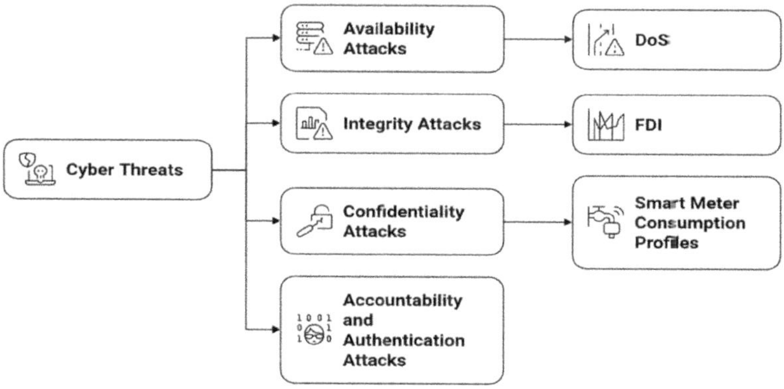

Figure 7.3 Cyber threats.

or manipulated, such as FDI. The third category is confidentiality attacks, where private data is leaked, such as smart meter consumption profiles. The fourth category is accountability and authentication attacks, where adversaries pretend to be legitimate users or hide their identity.

This simple taxonomy helps in understanding how wide the security problem is in smart grids. It is also important to see that cybersecurity issues in smart grids are not just theoretical but have a real-world impact. In Ukraine, hackers managed to take control of power distribution in 2015, which resulted in large-scale blackouts. In 2022 and 2023, researchers reported new vulnerabilities in solar panel control systems, which could allow attackers to shut down renewable power sources remotely. In 2024, several surveys pointed out that ransomware attacks on utilities are increasing because attackers know that electricity is a critical service that cannot be delayed. Utilities often end up paying ransom to restore services. These examples clearly show that the cybersecurity landscape of smart grids is becoming more complex every year. Smart grid cybersecurity therefore needs to be seen from multiple angles. It is not enough to only encrypt the data or install firewalls. Since the grid includes millions of devices, from very small smart meters to very large control centers, the solutions also have to be diverse. Some solutions must focus on lightweight cryptography for small devices; others must look at large-scale anomaly detection using machine learning (ML) in control centers. Regulatory bodies should enforce common standards across vendors, because if not, each vendor might implement security in their own way, and in doing so there are gaps that attackers can exploit. Thus, the smart grid cybersecurity landscape is very broad. The architecture itself is complex, and networking. Communication links, edge devices, data repositories, and central servers are the attack surfaces. The threat taxonomy reveals that attacks can be directed against availability, integrity, confidentiality, and authentication. Incidents in the real world prove that these threats are not a figment of our imagination but are actually occurring. All these facts confirm that cybersecurity and privacy in the smart grid cannot be assumed as an afterthought feature. It must be part and parcel of design, planning, and operating the modern energy grid.

7.3 PROBLEM DIMENSION ANALYSIS

7.3.1 Integrity and authentication failures

One of the non-negotiable requirements of a smart grid is that any data traveling from one location in the network to another must be trusted. If the control center receives measurement values from a smart meter or from a substation measurement sensor, it must ensure that the information is correct, and that it has not been altered along the line. This need is referred to as data integrity. Likewise, the control center must be sure of just who is feeding it data. If an attacker impersonates a smart meter and starts flinging out false values, the entire billing system collapses. This requirement is called authentication. In layman's terms, integrity just refers to the fact that the information is correct and has not been altered, and authentication refers to the fact that the data was sent by a legitimate sender. In the traditional grid, this problem was not too awful because there was not much communication. Power was transferred one way (from the plant to the consumer), and most devices were not even connected to any kind of extended network. The situation in the case of smart grids is different. Millions of smart meters are reporting data every 15 minutes. Substations send live status updates. Control commands continuously flow to switch devices on and off. All of this happens in real time and through open communication networks. This makes the system highly vulnerable to integrity and authentication failures. A common attack that targets integrity is the FDI attack. In this case, an attacker carefully injects false measurement data into the communication system. The false data is designed in such a way that it looks legitimate to the control algorithms. As a result, the control center makes wrong decisions. For example, if the load on a line is falsely reported as low, the control center may push more power into that line, which can cause overloading and failures. Similarly, if consumption is falsely reported as high, the utility may start unnecessary generation, waste fuel, and increasing costs. Authentication failures are also very serious. Imagine a hacker who gains control of a smart meter and then uses that identity to manipulate billing data. The hacker can reduce his own bill or increase someone else's bill. On a larger scale, if an attacker manages to impersonate a substation device, they can even send false commands to the grid.

These sorts of impersonation attacks can cause pandemonium. Several studies have shown that smart meters are being shipped with default passwords in poor condition or with outdated firmware and thus are easy to crack. There are existing countermeasures to these problems. One of the most popular answers is the deployment of digital signatures and public key infrastructure (PKI). Here, a digital certificate that proves the identity of each device is issued to each device. For any message sent by the device, it is signed with its private key, and the control center can check the signature with the corresponding public key. This way, while the data is consumed, there is no way for anyone to change it, and at the same time invalidating the signature. Another is the use of secure protocols such as transport layer security (TLS), often found over the internet to secure web traffic. By TLS, communication between smart meters and control center can be encrypted and authenticated. But even these solutions are not flawless. The most significant of those is scalability. "There are millions of devices and managing the digital certificates is not a simple task." Even if each smart meter needs to have its own certificate, then the utility would have to have a huge infrastructure for issuing certificates, updating them, and revoking them. This is further complicated when devices are distributed among varied vendors and jurisdictions. Another issue is key management. If an attacker steals a private key of a device, then the attacker can completely impersonate that device. Identifying and revoking compromised keys is a very challenging problem. Besides, some of the smart grid nodes are extremely small and not very powerful. Smart meters, for example, must be very cheap and power efficient. Running complex crypto such as RSA or Elliptic Curve on smart meters would be slow and can drain their resources.

This results in scenarios where the vendors either use less strong algorithms or, in some cases, do not authenticate at all, leaving additional avenues for vulnerabilities. Many recent studies have shown the seriousness of integrity and authentication failures. For instance, it has been demonstrated that a few measurements are sufficient for attackers to avoid detection and nevertheless to have significant effect on the grid. Another paper published in 2023 also observed that many utilities are still using outdated SCADA protocols that have a weak or no authentication and are therefore vulnerable to spoofing. These results only confirm that integrity and authentication are not only technical issues, but also management and policy issues. Overall, integrity and authentication failures are one of the most vulnerable parts of smart grid networks. They allow rogue data to be inserted by attackers and devices to be impersonated, which can lead to illegitimate decisions and blackouts. Despite some protection that is provided by digital signatures, PKI, and secure protocols, the scalability, device-centric key management, and device limitations remain. This makes integrity and authentication a high-priority area of research, engineering, and analysis for smart grid security-conscious researchers, engineers, and policy analysts.

7.3.2 Centralized architecture vulnerabilities

The design of most existing smart grid systems is based on a centralized model. In this model, a central control center acts as the brain of the entire network. All the data collected from smart meters, substations, and sensors are sent to this control center. The center then processes the information and sends back control commands. This design looks simple and efficient because everything is controlled from one place. However, at the same time, it creates a big weakness. If the central control fails, the whole system becomes useless. This is what is called the problem of centralized architecture vulnerabilities. In the old power grid, central control was not such a big issue. The grid was mostly static, and decisions were taken slowly. But in the smart grid, conditions change very fast. Renewable sources like solar and wind vary within minutes, and electric vehicles create sudden demand peaks. All this requires real-time monitoring and quick decisions from the control center. This heavy dependency makes the central servers a very attractive target for attackers. If an arbitrary is successful in shutting down or corrupting the central control, the whole smart grid may go down within seconds. One of the most popular attacks against centralized servers is a distributed denial of service (DDoS). During this attack, thousands of infected machines spam the control server with fake requests. The server becomes overloaded and cannot process genuine requests anymore. In a smart grid, this can be very harmful because even a small delay in control signals can destabilize the power balance. Another growing threat is ransomware. Attackers infect the central server with malicious software that locks important files or control programs. The attackers then demand money to unlock them. Since electricity is a critical service, many utilities feel forced to pay the ransom just to restore operations. Real-world cases show that such threats are not just theory. In 2021 and 2022, several utilities in North America and Europe reported ransomware attacks on their IT and operational systems. In many cases, the control centers were affected, leading to temporary shutdowns of energy delivery. A 2023 survey highlighted that centralized grid systems are particularly attractive to ransomware groups because they know that the downtime of electricity cannot be tolerated., This shows that centralization acts like a single point of failure, which adversaries can exploit. Another problem with centralization is insider threat. Since all the control is in one place, employees or contractors who have access to the control systems can cause large damage if they misuse their authority. For example, a disgruntled employee may change settings or leak sensitive information. In a decentralized system, the impact of such insider actions might be limited, but in a centralized system, the damage can spread across the whole network. Researchers have therefore started to explore decentralized alternatives. One approach is to use federated learning and edge computing. Instead of sending all the data to the central control, edge devices process data locally and only share the results. This moves the load off the centralized server and also makes the system less centralized for a single point of failure. Another method is through the use of blockchain, where transactions and operations are recorded on many nodes and not on a single server. This way, if one node is attacked, the other nodes can keep running safely. Decentralization though comes with its own set of problems as well. Edge devices are not suited for heavy computation. Energy demands vary from second to second, making it challenging to synchronize blockchain nodes in real time. Still, distributed systems are seen as a necessary step to avert the dangers of centralization. Governments and researchers are actively promoting hybrid models where some control remains centralized for efficiency, but critical operations are distributed for resilience. In conclusion, centralized architecture vulnerabilities represent a major security risk for the smart grid. A single attack on the central control can affect millions of consumers. Real incidents of ransomware and DoS show that this is a real and urgent problem. While decentralization by edge computing, federated learning, and blockchain is a potential direction, there are still open challenges. Balancing efficiency and resilience provides the solution. In order to be reliable and secure simultaneously, the smart grid needs to decentralize its dependency from a single central point as much as it is feasible.

7.3.3 Privacy limitations in blockchain

Blockchain has been widely discussed as a possible solution to many smart grid problems. The conceptualization of blockchain is that no data is stored in a single central server, yet it is stored in multiple nodes. Every transaction gets stored in blocks, and after a block gets validated, it becomes practically difficult to modify it. This makes it highly promising for smart grids where integrity, trust, and transparency matter. We can use blockchain for maintaining a record of energy trades by consumers, for smart meter reading verification, or for maintaining billing records. Since there is one common ledger shared by everyone, no one needs to trust a common authority. Figure 7.4 depicts various issues of smart grid security.

Yet at the same time, the radical openness of blockchain gives rise to a paradox of privacy. In most public blockchain systems, all the transaction data is visible to every participant. Even if the actual data is encrypted, the metadata such as the time of transaction, the size of transaction, and the address of the participants remains open. In a smart grid scenario, this can be a serious issue. If a user is selling or buying electricity using blockchain-based peer-to-peer trading, other participants may be able to infer the consumption pattern of that user. Over time, they can build a profile of when the person is at home, how much energy he uses, and even what appliances may be running.

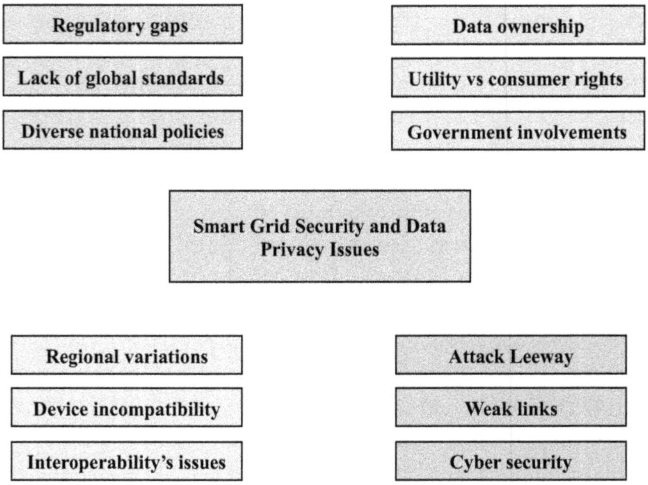

Figure 7.4 Challenges in smart grid.

Another limitation is that blockchain does not provide anonymity by default. Each user is identified by an address. While the address itself may look like a random string, with enough data, it can be linked back to the real identity of the user. In financial blockchains like Bitcoin, there are already many examples where transaction patterns have been used to trace identities. The same risk exists in smart grids. Once the address of a smart meter is linked to a household, all the energy usage information of that household becomes exposed. There have been proposals to use privacy-enhancing technologies on top of blockchain. One of the methods is zero knowledge proofs (ZKP). With ZKP, a user can prove that a transaction is valid without revealing the details of the transaction. Another method is to use ring signatures or mixing protocols, where multiple transactions are combined so that individual patterns are hidden. Homomorphic encryption is also being studied to allow computations on encrypted data without exposing the raw values. While these methods show promise, most of them are still in research phase and are not yet deployed in large-scale smart grids. Another issue is that introducing privacy features tends to add complexity and cut back on efficiency. Smart grids run in real time, and wait times of just a few seconds can introduce instability. If privacy mechanisms make blockchain operations slower, then they may not be practical for real-world use.

For instance, homomorphic encryption is safe, but it is very computationally intensive. Similarly, increasing latency can be caused by trying to mix transactions to hide identities. This leads to a privacy/performance trade-off. Recent research between 2022 and 2025 has defined this privacy paradox very clearly. Blockchain is viewed as an effective tool to achieve both trust and data integrity. On the other hand, it is transparent, and this leads to the leak of sensitive information. A 2023 research noted peer-to-peer energy trading via blockchain could actually amplify the risk of profiling attacks on consumers. The 2025 paper showed a hybrid blockchain encryption scheme with extra encodings, but also showed performance problems. These results demonstrate that blockchain cannot be uncritically applied in smart grids without thinking about privacy constraints. In summary, blockchain is a great technology that gives good integrity but poor privacy. The transactions that are open mean that users can be tracked and profiled. Although promising cryptographic techniques such as ZKP, ring signatures, homomorphic encryption, etc., can aid, they are still in the immature stage and suffer from a performance penalty. This presents us with an open problem for maintaining privacy for smart grids based on blockchains. There is a need to find the right balance between secrecy and free exchange of information; it is important for the engineers and researchers to ensure that blockchain applications can be applied in smart grids, but without endangering the privacy of consumers.

7.3.4 Resource constraints in edge devices

Smart grids are not comprised of large power stations and centralized control points only. Much of the grid has been replaced by very large numbers of small devices at the network edge. They are smart meters at homes, sensors on transformers, controllers on electric vehicle charging stations, and sensors on renewable plants. These are what can be referred to as edge devices. They are significant as they gather the raw data and transmit it to utilities, and in some cases, they make instant local decisions. The smart grid is unable to operate without edge devices. But there is a big problem. These edge devices are typically inexpensive, small and power-efficient. Millions of household customers cannot have a smart meter that is very expensive and will reject it. On the same note, the processing power and memory of field sensors cannot be very high lest they become very big and expensive. The net effect is resource-constrained devices. They are low-powered in terms of central processing unit (CPU) speed, they have little memory and have small storage. Due to this, it becomes extremely hard to impose stringent security systems on them. As an example, higher-order encryption, such as homomorphic encryption or lattice-based post-quantum encryption, is computationally intensive. The time to execute these algorithms on a smart meter is either too long or it may even fail due to inadequate memory. Such devices are capable of slowing down even simple algorithms such as Rivest–Shamir–Adleman (RSA), when used thousands of times a day. This makes vendors tend to use weaker algorithms or in some cases, they do not bother using encryption at all. This renders the edge as the most pathetic node in the entire system. Attackers are aware of this and they tend to compromise smart meters or sensors as points of entry into the grid [13]. Routine updates are another problem. There is no problem in pushing security updates or patches whenever a new vulnerability is discovered in large servers. This is not so easy in the case of edge devices. The number of smart meters that are installed in the field cannot be massively updated in a short time. In some cases, equipment is put in remote locations, where physical access is hard. This implies that a number of devices have been operating with out-of-date firmware for many years. Attackers would utilize the old vulnerabilities that have not been patched. This has been attempted to be solved by cryptographic scholars in the development of lightweight cryptographic algorithms. These algorithms provide a security/efficiency trade-off. As an example, elliptic curve cryptography (ECC) is not heavy, like RSA and is being used in miniature devices. Similarly, some block ciphers have low memory needs. Nevertheless, in spite of these enhancements, there is still a difference between robust security and capacity in the devices. There are proposals to utilize hardware-assisted encryption, in which cryptographic functions are directly implemented in the chip. This is able to provide powerful protection without being a load to the processor. In the last 2022–2025 studies, the risk of resource-constrained devices has been mentioned multiple times. In 2023, a survey indicated that a large number of deployed smart meters continue to use obsolete cryptographic criteria. In 2024, a paper on lightweight lattice encryption suggested new algorithms, but also acknowledged that it will be slow to adopt, because of cost and integration challenges. Such reports prove the fact that edge devices are now the weakest link of the smart grid. When the edge is weak, then, regardless of the security of the central systems, attackers will still find their way to the systems. Finally, one of the greatest challenges to smart grid cybersecurity is the constraint of resources in edge devices. The machines are too small to compute intensive algorithms, too numerous to be updated frequently, and too feeble to protect themselves. Lightweight and hardware-based solutions provide some relief, but the problem is not solved yet. To have a really secure smart grid, we have to make the protection methods of these small edge devices powerful and practical. Otherwise, the target, the weak point of attackers is always on the soft side.

7.3.5 Regulatory and interoperability gaps

A smart grid is a socio-technical system, besides being just a technical system. This is to say that it is not only influenced by devices and algorithms but also laws, regulations, and standards. Policies regarding the way energy is to be managed, the manner in which consumer information is to be processed, and the manner in which utilities are to be managed are different in different nations. Theoretically, the said policies are meant to render the grid secure and fair. However, regulatory gaps and interoperability gaps are much more challenging data privacy and cybersecurity issues in practice. One of the biggest challenges is that there is still not one global standard for smart grid security. General Data Protection Regulation (GDPR) stresses the importance of privacy in Europe, and conducts measures to ensure that customers' data are collected and stored responsibly. Standards in North America like NERC CIP include policies that are geared toward the protection of critical infrastructure, where monitoring and reporting are strict policies. There are some guidelines in piecemeal countries of Asia. Because of this diversity, smart meters/sensors vendors usually have different regulations per region. If such devices are interconnected in a large grid, they may not behave the same. As a consequence, this interoperability disadvantage provides attackers with a leeway to attack the weakest link in the system. The other gap is related to data ownership. Once consumption data has been collected using smart meters, then the utility, government, or consumer owns that data. To date, this question is still beyond answer in most quarters.

Ownership can be claimed by utilities to optimize and analyze the grid. Policy-making may be demanded by governments. Consumers might desire having control to safeguard privacy. Lack of a proper regulating structure gives rise to confusion and in some cases legal battles. In the meantime, the hackers may exploit this ambiguity, stealing information without being held very strongly accountable. It is also becoming a factor in cross-border energy trading as renewable energy is expanding. There is an increased linking of grids between countries to equalize the energy requirements. However, when electricity is transferred across borders, the information is also transferred across borders. When the level of security is low in one of the countries, the entire system is exposed. The need to harmonize energy data policies in the European Union has been discussed already. In the absence of such harmonization, the attackers can use the variance in country standards [14]. The other thing of significance is compliance. Standards are not always applied in all utilities, even when they are set. Smaller utilities might not have the means to use complicated security structures. Others will take time to comply since it will be costly. As a result, the level of protection varies widely across the grid. Inconsistency in compliance presents attackers with the mere slightest attack point. These regulatory and interoperability gaps have also been noticed in recent studies. One 2023 study wrote that technical solutions, like blockchain or federated learning, are developing but the policy and legal component lags [15]. One 2024 survey noted that non-uniform regulations are also emerging as a cause of uncertainty among vendors and as a challenge to developing general-purpose security solutions.

In a 2025 paper, it was warned that unless international cooperation is organized, the vision of an international smart grid achieved by global security could not be fully realized. Briefly, the smart grid cannot be made secure by technical improvements only. The system will be vulnerable unless it is regulated, with proper ownership policies, harmonized cross-border standards, and stringent adherence. The attackers are indifferent to the national boundaries or the jurisdictional matters. They will eternally take advantage of the weak point. Thus, the regulatory and interoperability gaps need to be closed as much as possible, as the technical issues in smart grids are.

7.4 STATE-OF-THE-ART SOLUTIONS

Cybersecurity and privacy of smart grids are the issues that have drawn the attention of numerous studies during recent years. Efforts have been made in trying to come up with solutions by researchers, governments, and industries. They are some technical, some policy-based, and some hybrid in nature. During 2022–2025, numerous papers and reports were introduced with new concepts of how the smart grid can be secured against various categories of threats. In this section, we are going to touch upon some of the principal solutions that are under consideration today.

7.4.1 Blockchain-based privacy frameworks

Among the most popular directions has been the usage of blockchain technology. As stated above, blockchain offers a distributed registry in which every transaction is stored in a transparent and secure way. Scientists have attempted to apply blockchain to such areas as peer-to-peer energy trading, billing security, and trust in the smart grid. As an illustration, a 2023 article suggested a blockchain-anchored billing arrangement in which consumers have the potential to exchange electricity without addressing a central body. It was also suggested to use blockchain together with homomorphic encryption to enhance privacy in energy transactions in another 2025 paper. The benefit of blockchain is that it does not require the reliance on a central control, due to which the system becomes less vulnerable to centralized attacks. It also makes sure that a transaction cannot be easily changed when it is recorded, which enhances integrity. Nevertheless, as observed in Section 7.3, the key issue with blockchain is the leak of privacy. In response to this, researchers are attempting to add additional privacy layers such as ZKPs and mixing protocols. There are prototypes that have a good result, but the majority are still in early testing phases.

7.4.2 AI and ML-based intrusion detection

It is also observed that AI and ML are viewed widely as relevant tools to enhance security in smart grids. The reason is simple. A smart grid of the present time produces an immense volume of data in a second. Information is continuously sent by smart meters, sensors, control systems, substations, and even electric vehicles. Human operators cannot possibly manually examine all this data to determine abnormality in it. This is where AI and ML can be of great use since they can automatically notice abnormal behavior that can signal an attack. The key concept of AI use is that the system will be able to learn the normal and then recognize the abnormal occurrences. As an illustration, when a smart meter shows an unreasonably high consumption at midnight, the ML model might be raised as suspicious. Equally, when a substation is transmitting control signals using an unusual frequency, the system can readily initiate an alarm. This type of automatic pattern identification renders AI appealing for intrusion detection in smart grids. In 2022–2025, different methods of ML have been tried. There were those works where neural networks were used, whereas there were also those works where support vector machines and ensemble models were applied [16]. These algorithms are fed with normal operations history. When live data arrives later on, the model developed learned patterns with which it compares the data. When a mismatch is already found in the newly obtained data, the system suspects intrusion. It was demonstrated in a 2023 paper that deep learning models are more accurate at detecting FDI attacks than the traditional methods. Ensemble learning was applied in a 2024 study to the detection of DoS attempts in communication to the grid. These articles demonstrate that indeed, AI can enhance the intrusions detection speed and accuracy. Nevertheless, significant problems also exist. Even ML models can be deceived. Attackers may target the data by slightly modifying the data to make the model give an incorrect decision in what is known as adversarial attacks [17]. Another problem is data poisoning. This is where training data is important, as when a machine is fed with the wrong idea, the ML model learns the wrong way and cannot identify actual attacks. And so, while AI is going to be very beneficial, it is not going to be the silver bullet. It must be supported by other security measures, like encryption, authentication, and human vigilance. Irrespective of the negative aspects, ML and AI technologies are a highly needed element of the smart grid security. They are more accurate than rule-based systems, easier to deploy to emerging attack patterns, and provide real-time detection. The current research agenda is to create such models to be more robust, explainable, and resistant to adversarial manipulation. Over the next several years, it is probably that AI-based intrusion detection will form a central part of the smart grid defense architecture and assist with cryptography methods and regulations to create a more secure world.

7.4.3 Lightweight cryptography for edge devices

As already mentioned, edge devices such as smart meters, IoT sensors, and home energy controllers are extremely resource-constrained. They are miniature devices that are intended to primarily measure energy use and report to the utility. These devices lack powerful processors and huge amounts of memory as compared to laptops and servers. Meanwhile, they have a direct connection to the grid network, so they become a target of attackers. Without the adequate security of these devices, they can be hacked, then utilized to steal their data or even to disrupt the grid. Because of this fact, scientists have been looking for cryptographic techniques that can provide security without compromising the tiny hardware. This is where lightweight cryptography has come into the picture. Conventional algorithms such as RSA or a full-size advanced encryption standard (AES) are very secure but also very heavy [18]. RSA in particular requires very large size keys in order to be secure. The process of running such kinds of algorithms on a smart meter can slow down its normal operation and even lead to faster battery depletion. For example, if a meter had to sign every reading using RSA, then that would consume too much power and probably not be practical. To remedy these problems, scientists have looked at smaller and faster algorithms that still provide a reasonable level of protection. One of such options is ECC, which requires fewer keys and yet is as secure as RSA. Practically, a 256-bit ECC key is equivalent to a 3072-bit RSA key. This implies that the computation is significantly less heavy, and is appropriate in low-CPU devices. As a matter of fact, several prototypes of smart grids between 2022 and 2024 have tried ECC to authenticate smart meters. The findings demonstrated that ECC might be utilized to sign meter data with significantly reduced latency than RSA and was therefore more feasible in real-time grid communication [19]. The other interesting area of study has been lattice-based lightweight cryptography. This form of encryption is believed to be resistant to quantum attacks, something significant because in the future, quantum computers can be capable of cracking RSA or ECC. The lattice-based schemes tend to be heavy; in some works, simplified schemes have been proposed, and still can be applied in constrained devices. One 2023 study showed a lattice-based protocol on smart meters that could be resistant to quantum threats and still within the constrained memory. This demonstrates that lightweight cryptography does not only concern the present, but also the future. Along with the software-based solutions, hardware-assisted security is also placed in a lot of emphasis [20]. Encryption capabilities are embedded in the microchips of the smart meters in this method. Rather than using software libraries alone, the hardware itself contains secure modules, where encryption, decryption, and key storage are accomplished. This has two main benefits. One is that it decreases the processing load on the central chip since the security functions are offloaded to specialized circuitry. Second, it enhances the system with the keys being stored in hardware that is significantly more difficult to steal the keys. One 2024 paper proposed hardware security modules (HSMs) within smart meters. These HSMs store the keys in a secure location and encrypt them in a secure place. The thief can not steal the keys by stealing the hardware easily, even in the event of the main device being breached. This solution is not new in the form of payment cards and in safe mobile devices and therefore, extending it to the situation with smart meters is the logical step. But the challenge is cost. Millions of meters of HSMs added will increment the total cost, and utilities are typically highly sensitive to cost when large-scale deployment is required. Another impediment is compatibility. The lightweight algorithms that are selected may vary among different vendors. One machine can use ECC, another PRESENT, and a third can come up with a hardware design of another kind of its own. Incompatibility occurs when the devices with different approaches to communicating with each other seek to communicate. That brings about slow uptake on a mass scale. A certain amount of world standardization is in progress that standardizes common lightweight cryptographic approaches between the IoT and the devices of the smart grid, but no consensus remains absolute. Despite all these difficulties, the need to have lightweight cryptography is inevitable. Edge devices are the weakest point of the smart grid, and there is always a way the attackers can exploit it once it is vulnerable. A track where minuscule machines obtain a satisfactory level of protection without affecting their main objective can be found in lightweight cryptography. The dilemma of strength and efficiency is the most important and, in many cases, a difficult one. Lighter and algorithm are susceptible to breaking. It cannot be done by heavier and the device. Put simply, edge device lightweight cryptography is the continuing process of bringing security to life in the real world. It is not a search about finding the perfect protection but rather a search to make the billions of small devices within the energy ecosystem at least able to protect themselves against the common attacks. In the coming years, as further research streamlines these techniques and as standards get clearer, lightweight cryptography will probably become a standard feature of all smart meters and edge devices of the smart grid.

7.4.4 Federated learning for distributed anomaly detection

Federated learning is another new concept that has been applied to smart grid security. In federated learning, instead of sending all the raw data to a central server, the devices train local models and only share the model updates. The central system then combines these updates to form a global model. This way, sensitive data does not leave the device, which improves privacy. In the context of smart grids, federated learning can be used for anomaly detection, demand forecasting, and secure energy trading. A 2024 survey discussed the use of federated learning for privacy-preserving load forecasting and intrusion detection. Now as the consumer data is present locally, so privacy is better protected. But federated learning also faces challenges like communication overhead, heterogeneity of devices and vulnerability to poisoning attacks.

7.4.5 Regulatory proposals and compliance models

Apart from technical solutions, researchers and governments have also focused on policy and regulation. Many reports between 2022 and 2025 have stressed that without clear regulations, even the best technical solutions may not work effectively. For instance, the European Union's GDPR places strict regulations on the protection of consumer data. In North America, critical infrastructure security is the subject of NERC CIP standards [21]. Several papers have proposed the development of international standards for smart grid data privacy, in particular for cross-border electricity trading. There are also talks of compliance models. These are frameworks to help utilities look at the question of whether they are conforming to the required standards. An automated compliance auditing system for smart grids was proposed in a 2023 study that was able to continuously test devices and systems to see if they are meeting security requirements. This is important because smaller utilities are often targeted by attackers due to weak compliance.

In summary, state-of-the-art solutions for smart grid cybersecurity and privacy cover a wide range of approaches—blockchain for distributed trust, intrusion detection by AI, lightweight cryptography protecting small devices, privacy-preserving models by federated learning, and regulations defining the policy environment. It is interesting that all these solutions are promising, none of them perfect, which means they all have advantages and disadvantages. So, the issue lies in combining them in a way that creates a secure, private, and reliable smart grid.

7.5 OPEN RESEARCH CHALLENGES

Even though a lot of progress has been made in smart grid cybersecurity and privacy between 2022 and 2025, the field is still full of challenges. Researchers have proposed many solutions like blockchain, federated learning, lightweight cryptography, and intrusion detection systems, but when it comes to large-scale practical deployment, many gaps remain. The smart grid is a very complex environment that includes millions of devices, multiple vendors, different communication protocols, and cross-border regulations, and due to this complexity, no single solution can solve all the problems. In this section, we will highlight the main research challenges that still need attention.

7.5.1 Balancing security with real-time performance

One of the uphill tasks is to balance strong security with the need for real-time performance. A smart grid requires decisions in milliseconds. For example, a sudden fault in a substation will require the control system to react instantly to prevent a blackout. If the security mechanisms are too heavy, they may introduce delays. Encryption, authentication, and anomaly detection all require computation, which may slow down the system. Homomorphic encryption and blockchain are good examples. They provide strong integrity and privacy, but are computationally expensive. Running them in real-time energy markets or grid operations may not be practical. Researchers are trying to design lightweight versions, but the trade-off between performance and security remains open.

7.5.2 Scalability of security mechanisms

Another major challenge is scalability. Most solutions look good in small-scale pilot projects but fail when millions of devices are involved. PKI is one such example. In theory, digital signatures and certificates can authenticate devices perfectly. But when each household smart meter requires a certificate, the utility has to manage millions of keys, renew them, and revoke them in case of compromise. This becomes almost impossible without massive overhead. Blockchain also suffers from scalability issues. The more transactions are added, the bigger the ledger grows. In a large smart grid with millions of micro-transactions every day, blockchain ledgers can become too heavy for small devices. Therefore, how to scale security solutions to national or even continental level grids is a big research challenge.

7.5.3 Privacy preservation beyond data encryption

Encrypting the data is not enough for a smart grid's privacy. Even if the consumption data is indeed encrypted, adversaries can learn patterns from metadata. That is, for a smart meter, the rate and time of message transmission can reveal the time a home is in operation, when it is inhabited by the occupants, and even what appliance activity is occurring. This is known as the problem of inference attacks. Researchers have proposed differential privacy, federated learning, and data anonymization to counter such attacks, but each of these has drawbacks. Differential privacy adds noise, which can reduce the accuracy of energy forecasting. Federated learning protects raw data but may still leak information through model updates. Therefore, to design privacy-preserving methods that maintain data utility is still an open question.

7.5.4 Resource constraints in edge devices

As explained earlier, edge devices like smart meters are the weakest link in the chain. They are cheap, small, and resource-constrained. Running heavy cryptography on them is very difficult. Even lightweight algorithms can sometimes be too much if the devices also need to perform measurement and communication tasks. The challenge is not just computational.

Updating firmware is also a big issue. Millions of devices are deployed in the field for more than 10 years. If a new vulnerability is found, updating all devices quickly is very hard. Attackers often exploit this weakness. Research is needed to design self-healing, auto-updatable, and lightweight secure devices that can work in constrained environments.

7.5.5 Interoperability and standardization

Another challenge is interoperability. Different vendors design devices with different security protocols. One smart meter may use TLS, another may use proprietary encryption, and a third may use no encryption at all. When these devices communicate in the same grid, compatibility issues arise. Attackers can take advantage of the weakest protocol. Standardization is supposed to solve this problem, but currently, global standards are fragmented. Europe, the United States, and Asia all follow different regulations. Without harmonized standards, it is difficult to implement uniform security. Research is needed on creating universal frameworks that can adapt to different regions but still maintain baseline security.

7.5.6 AI and ML vulnerabilities

ML-based intrusion detection has become very popular. But ML models themselves are vulnerable. Attackers can perform adversarial attacks by slightly modifying data so that the model makes wrong predictions. For example, by carefully altering meter readings, an attacker can bypass an intrusion detection system. Another risk is data poisoning. If attackers inject false data during the training phase, the ML model will learn wrong behavior. This means that while AI improves detection, it also introduces new attack surfaces. Research is needed on making AI models more robust, explainable, and resilient against adversarial and poisoning attacks.

7.5.7 Resilience against evolving threats

Cyber threats are not static in nature; therefore, attackers constantly refine their techniques because what works today may not work tomorrow. For example, ransomware was not considered a major risk for smart grids a few years ago, but today it is one of the biggest threats, with multiple real-world cases reported. Similarly, quantum computing is expected to take over and breach classical encryption in the near future and if smart grids continue to use RSA or ECC, then the entire system will be at risk. Therefore, research is needed in post-quantum cryptography and adaptive defence systems that can evolve as threats change. This requires continuous monitoring, updating, and upgrading of security systems, which is still a major gap.

7.5.8 Regulatory and policy challenges

Technical solutions alone cannot solve the cybersecurity problem. Regulatory and policy frameworks are equally important. Today, regulatory standards vary widely across countries, where some countries focus heavily on privacy, while others prioritize critical infrastructure protection. This means that without international cooperation, attackers can exploit weak spots. Another issue is compliance. Even when standards exist, not all utilities follow them strictly. Smaller utilities may lack resources to implement complex frameworks. Research is needed on low-cost compliance models and automated auditing systems that make it easier for small and medium utilities to meet security requirements.

7.5.9 Human and organizational factors

Finally, one of the most neglected challenges is the human factor, where employees, contractors, and even consumers can unintentionally cause security problems, which include weak passwords, careless handling of data, phishing emails, and lack of awareness, all of which contribute to vulnerabilities. In the study from 2022 to 2025, the reports have demonstrated that many attacks have succeed not because of technical flaws but because of human errors. Therefore, cybersecurity training, awareness campaigns, and organizational policies are essential. Research is needed on designing human-centered approaches that integrate technical solutions with social practices.

7.5.10 Integration of multiple solutions

The last major challenge is how to integrate multiple solutions together. Blockchain, AI, lightweight cryptography, federated learning, and regulations possess individual strengths. But when individual components are introduced independently, possibly one might interfere with another. By introducing blockchain, for instance, one might enhance transparency while degrading privacy. By using strong cryptography, one might boost security while degrading performance. Therefore, the challenge is creating holistic, multifunctional solutions through balanced integration of multiple methods, which requires collaboration between computer scientists, electrical engineers, policy administrators, and industry. Otherwise, the smart grid will always be an ad hoc collection of local solutions. Briefly, open smart grid cybersecurity and privacy challenges are multi-faceted, which encompass technical challenges such as scalability, performance, and edge constraints and policy and human challenges such as regulation, compliance and awareness. The challenges are interrelated and addressing one or some of them without reference to the rest may not be useful [22]. So,

Smart grid future security is based on solving open challenges holistically and collectively.

7.6 FUTURE RESEARCH DIRECTIONS

The development of smart grids is progressing at an extremely rapid pace, and so are the attacks on these systems. The current solutions such as blockchain, AI intrusion detection, lightweight cryptography, and federated learning are helpful solutions, yet will not suffice in the long term. With increasingly larger and complex grids, we must consider future research directions that will offer greater security and privacy and be performance-wise feasible. One of the promising directions based on existing trends and gaps is indicated in the following subsections.

7.6.1 Integration of ZKPs and blockchain for privacy-preserving energy trading

The future of smart grid security is one of the most promising research directions that involves ZKPs and blockchain, which allow trading energy in a privacy-preserving manner. This is currently shifting energy trading toward a less centralized trade model. Due to the development of renewable energy sources like small wind generators and solar generators, consumers are gradually turning into producers, otherwise known as prosumers. These prosumers would prefer to resell the excess to the neighbors or to the main supply network. Blockchain has been viewed as a worthy contender to facilitate such peer-to-peer trading since it is capable of delivering tamper-proof records and eliminates the necessity of a central authority. However, even the very transparency that draws blockchain appeal generates a privacy issue. All the transactions registered on a blockchain are public to all the participants. Although the amount of power traded may be encrypted or obfuscated, the very fact that one house is busy so often and another so rare often can reveal patterns in the actions of the users. As an example, a criminal individual may read blockchain data and determine, at least approximately, when a family is home, when they use the power-intensive appliances, or when they are going on vacation. This information leakage is highly dangerous to the privacy of the users. This is the place where ZKPs are involved. A ZKP can be used by an individual to demonstrate the fact that something is true without any additional information.

The simplest example will be the situation when a prosumer wants to demonstrate that they have sufficient energy credits to sell. Using ZKP, the prosumer can demonstrate this fact without disclosing the exact number of credits. What is more, the smart meter might certify that the meter is in a valid range but leave the exact range secret. This way, the secret information is not leaked, and the trade system is guaranteed that the transaction is authentic. Blockchain and ZKP make the best pair of the public and private. Blockchain ensures that records will never be tampered with again, and ZKP ensures that personal information is never revealed where it does not need to be revealed. Collectively, the two can ensure energy exchange becomes reliable and private. This balance is crucial since players are not willing to trade when there is no reliability, and players are likely to feel vulnerable and exposed when there is no confidentiality. Researchers already began to consider this integration between 2022 and 2025. As an example, a 2023 article made a proposal for a ZKP-based blockchain in a peer-to-peer energy market. The design indicated that the participants were able to buy and sell energy units and conceal their consumption to others. The other article of the year 2024 was based on the efficiency of ZKP in the environment where the commonality of producing the evidence and the checking of it would not cause long delays. These affirm the potential of the idea but cite the difficulties that would be faced. Scalability is the first big challenge. Millions of smart meters may exist in a national smart grid, and there may be millions of energy trades in a day. It is not simple to generate and verify ZKPs on such a large number of transactions. Blockchain has its own scaling issues, but adding ZKP only increases its weight. In the event that the system may be too slow, it will not be practical in real-time trading of energy, where transactions have to be cleared within a short time. Thus, to scale to large-scale grids, future research should consider optimizing ZKP algorithms, perhaps with hardware acceleration or parallel processing. The second problem is resource limitations on edge equipment. IoT devices and smart meters are extremely low processing power devices. ZKP algorithms can very easily exhaust the capacity of such small machines. To make ZKP practical, researchers must find light variants or divide some portion of the generation of the proofs to a slightly more powerful local hub. Otherwise, the edge devices will be the limiting point of the system. Interoperability is another issue. Various blockchain systems can adopt different ZKP schemes. When multiple vendors implement their own, then inter-regional or inter-vendor integration will be complex. The factor of cost is also significant. The adoption of blockchain and ZKP does not only need new software but potentially may also need new hardware, training, and system design. The utilities are conservative in nature and are slow to take up technologies that might be costly to invest in. When the cost–benefit is not apparent, utilization may be postponed. The future research should, therefore, also be concerned with the optimization of costs and real-life benefits expressed in pilot projects. Nonetheless, blockchain and ZKP integration have a very good future. It is directly related to one of the dilemmas of smart grid security, which concerns the tension between transparency and privacy. Having blockchain only brings trust without privacy. With the addition of ZKP you get both. Moving to the future, a research area will be to create layered systems. Blockchain is responsible for recording and consensus and in a design like this ZKP can be applied selectively to the most sensitive transactions. The other path is to investigate hardware-assisted ZKP, in which cryptographic proofs are produced in less time by means of special chips installed in smart meters. The paper also entails hybrid solutions with small-scale microgrids of a territory using lightweight ZKP protocols and the bigger scale regional grids use more sophisticated ones with gateways assuming the translation role. To sum up, ZKP and blockchain in the field of privacy-preserving energy exchange are not a theory, but must be implemented in practice. As the number of prosumers increases and the frequency of decentralized exchange becomes more frequent, the need to have privacy would increase. In the absence of it, the users will not be keen to connect and this will become a bane to the future of the smart grids. The combination of the unchangeability of blockchain and the anonymity of ZKP will enable the smart grids of the future to create a trusted and secured exchange arena in which the players will feel empowered yet safe. This places the ZKP-blockchain integration among the most important research areas of the coming decade.

7.6.2 AI-driven adaptive authentication for real-time integrity

One of the simplest elements of cybersecurity in smart grids is authentication. It entails authentication of the device, or the user transmitting data that they are exactly who they say they are. Attackers can impersonate a smart meter or a control centre with ease without authentication and transmit false information. However, authentication is usually fixed in the present smart grid systems. After a device is registered and assigned a key or password, it proceeds to use the same means for months or years. This poses a serious weakness since modern-day attackers are not the same. They continue to switch methods; they can steal keys, make assumptions based on credentials, or even hack the communication channel. In order to address the issue, scientists are shifting toward AI-based adaptive authentication. The point is that, authentication must be dynamic and must vary depending on the scenario. As is the case with human beings, when we come across somebody that we have never seen, and who came to our door, we might first question the person before we trust them. AI is capable of being trained on the normal behavior of smart grid devices. In this case, a typical household smart meter transmits data at 15-minute intervals with deviations. When the same meter has suddenly begun to transmit data at one-second intervals, that is abnormal. This anomaly can be identified in real time with the help of AI. After it is detected, the requirement of authentication can be adjusted by the system. It might also require a more robust cryptographic verification than before or even a multi-factor authentication of the data before it can be allowed into the control system, instead of the traditional key [23]. This dynamic method is potent in that it increases the difficulty of attacks. In a static system, there is awareness of attackers of what authentication is being used and the attackers can plan to the same. However, they are not able to know the degree of verification that they will encounter in an adaptive system. Even this uncertainty is a defence. In 2022 and 2024, a few research papers have already shown ML-based anomaly-based authentication. Such works demonstrated that even simple AI models could identify abnormal behaviour of a device with a high level of accuracy and adjust the security requirements dynamically. Nonetheless, it is not so simple to construct such a system. Real-time performance is one of the obstacles. Smart grid is concerned with serious operations that do not support prolonged delays. In case the AI models are too slow to process behavior, it might introduce delays in the authentication procedure. As an example, protective relays must respond in milliseconds in case of a fault. Authentication may even result in blackouts when there is excessive overhead. Consequently, scientists will need to come up with small AI models that can run in real time on limited devices. False positives are another issue. AI models do not always work. In some cases, they can consider normal behavior to be abnormal. If this happens all the time, devices can be subjected to stronger authentication every single time, and this could create frustration and unnecessary delays. However, in the case of false negatives, the abnormal behavior would go undetected. Finding an appropriate balance in this accuracy is also a big research topic. There is also a new threat from attacks against AI called adversarial attacks. This means that the AI models should be impervious to adversarial samples. This is an issue that is still in the initial research stages with respect to smart grids. Scalability, by the way where there are millions of smart meters and thousands of control devices that may be a part of a national grid. It is not easy to train and maintain AI models on all of them, and some researchers propose federated learning mechanisms where individual devices train locally and only transmit model changes, but this itself will be an issue when it comes to privacy and communications overheads. However, the potential of the AI-powered adaptive authentication is massive [24–26]. There could be no better time when it comes to the dynamism of threats operating in the sphere of cyberspace. Unlike fixed passwords or keys, adaptive systems can be dynamic, based on time and situation. As an example, a smart meter can use lightweight authentication in normal operation. However, full cryptographic signatures might have to be used in a possible attack. Flexibility: the grid is not held up all the time [27].

The next generation of research needs to work on the development of effective anomaly detection models that can be deployed directly to the edge devices with minimal resources. Hybrid techniques can also be potentially successful, such as small models that run on-site to respond quickly and bigger models that are used at control centers to conduct deeper analysis. Explainability is another area - grid operators have to know why the AI raised the level of authentication. In the absence of this, they might not trust the decision of AI. Adaptive authentication could thus be more palatable in the field through explainable AI methods [28].

7.6.3 Hybrid architectures

Risk of single point of failure is one of the critical issues in the current design of smart grids. Most of the locations tend to have a high degree of centralization in the system. The control centers located in the regions, such as the ones mentioned above, compile all the information of the devices, meters, and substations it is expected to operate. It makes the required decisions and transmits the instructions of control back to the devices once the information is processed in it, which makes it easy to manage, but not without a weak point. Attacking or breaking down of the control center can lead to the disruption of the whole region. This is known to attackers, and they tend to attack such central nodes. In order to overcome this, scientists are currently considering hybrid architectures. It is meant to unify the power of centralized control, edge computing, and distributed systems based on blockchains. Not even a single node of such an architecture is responsible for anything in particular. Rather, there is cross-functional sharing of responsibilities. The core layer will remain, as full decentralization might not be feasible in a national grid [29]. The key systems are able to offer coordination, policing, and large-scale optimization. They may, among other things, control total energy in regional terms or inter-country trading in energy. Yet not every decision is supposed to go through the central node. The edge layer is valuable in the local decision-making. The microgrids and edge devices can process information locally and make real-time responses [30]. In the case of edge devices, say, the system does not need to wait until the central system transmits remedial instructions in the face of an overload in the transformer. There is less delay and the system becomes fault-tolerant. Edge devices may also remain in operation even when the central system is not online at certain times. A third robustness level is added by the blockchain layer. Using blockchain, valuable exchange of data like energy transactions, meter reads, or device authentication can be stored in a distributed registry. The ledger is stored on numerous nodes, and hence it does not suffer a loss on the occurrence of failure or attack of a node since the knowledge is valid and secure [31]. Another application of blockchain in establishing trust among different parties is that it does not need a central entity to check everything. These three layers, central, edge, and blockchain, provide a balance. Central control, edge computing, and blockchain provide efficiency, speed, self-sufficiency, and resiliency, respectively (Figure 7.5) [32].

All put together, they render the system invulnerable to the point of failure as there is no single node that will go down and take the whole system with it. There are, however, difficulties in the construction of hybrid architectures.

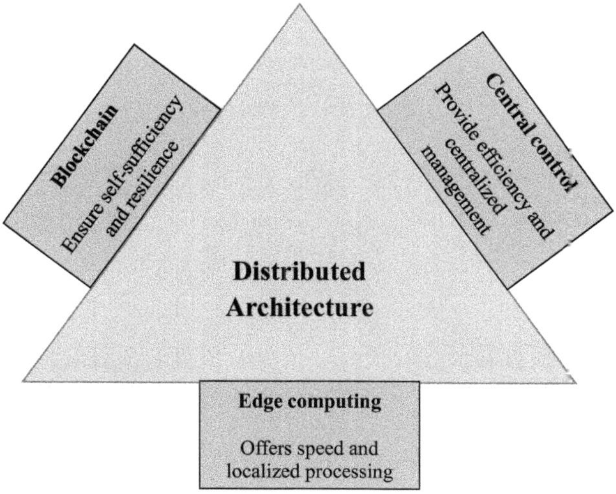

Figure 7.5 Blockchain distributed architecture.

The initial difficulty is coordination. Coordination of the decisions of central and edge systems is necessary. When the edge performs one action and the central system performs another, the conflicts can arise. The other difficulty is latency. Blockchain is not very easy to use as a time-sensitive control; its operations are slower than those of traditional databases. Lightweight or permissioned blockchains are being researched by the researchers and can be faster. Irrespective of these challenges, hybrid architectures are believed to be the future of smart grid design. They also find inspiration on the internet itself, which is hardened as it is distributed, not depending on a single server. As cyber threats keep on increasing, smart grids cannot afford to be dependent on central points. A hybrid overlay provides a way to be resilient, reliable, and secure.

7.6.4 Post-quantum security adoption in smart grids

Another major research direction is preparing for the quantum era because quantum computers, once their true potential is recognized, can crack traditional cryptography like RSA and ECC. Since smart grids are critical infrastructure, they cannot afford such a risk. The devices installed today may remain in operation for 10 to 15 years. By that time, quantum computers may already be available to attackers (Figure 7.6).

This means that we must start adopting post-quantum cryptography now. Post-quantum algorithms based on lattices, hash functions, and code-based systems are being standardized by organizations like National Institute of Standards and Technology (NIST). Some lightweight versions

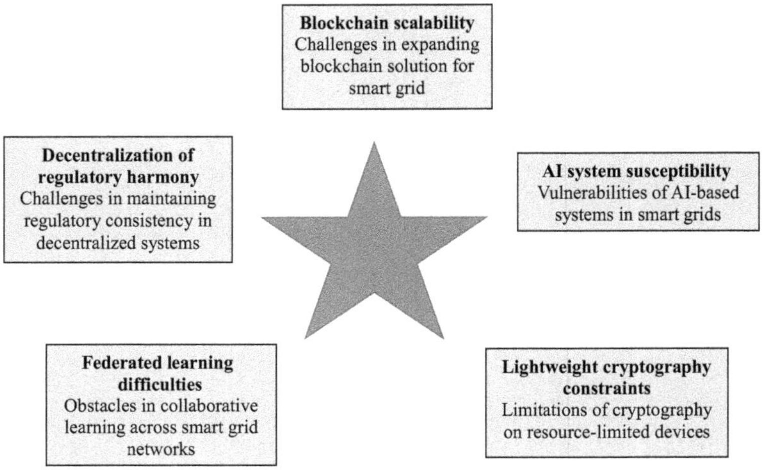

Figure 7.6 Smart grid security challenge.

of these algorithms are also being designed to fit into constrained devices. For smart grids, the challenge is to integrate post-quantum cryptography without overwhelming the small edge devices. A few studies between 2023 and 2024 tested lattice-based encryption for smart meters, showing that it can be done but with careful optimization. Future research should focus on making post-quantum schemes more lightweight and compatible with current protocols. Migration strategies also need to be developed because it is not possible to replace all devices overnight. Step-by-step upgrading while maintaining backward compatibility will be a big research and engineering task.

7.6.5 Global privacy-preserving compliance framework for interoperability

When we talk about cybersecurity in smart grids, it is not only about technology. A big part of the challenge comes from regulations and policies. Every country has its own set of rules for privacy and energy systems. For example, in Europe, the GDPR rules are very strict about how personal data is collected and used. In the United States, the NERC CIP standards are followed, which are more focused on protecting critical infrastructure. In Asia, there are different sets of fragmented laws depending on the country. Now the problem is that smart grids are no longer limited within one country. With renewable energy and cross-border electricity trading becoming common, these different laws create a lot of confusion and interoperability problems [33].

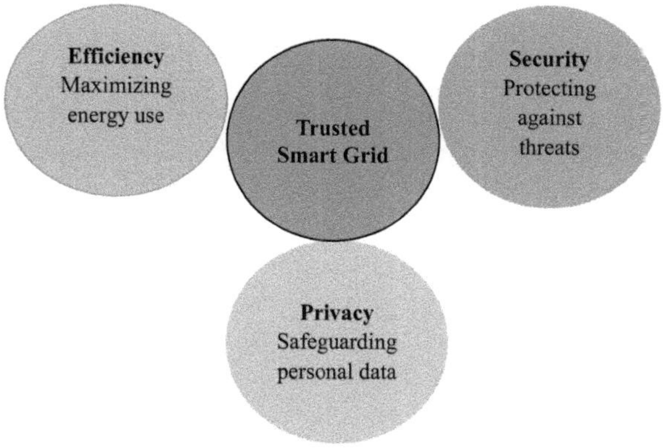

Figure 7.7 Smart grid equilibrium.

The really needed is a common privacy-preserving compliance framework that works at the global level. Such a framework should clearly state the basic principles of how consumer data should be handled, who owns the data, and who is accountable if something goes wrong. At the same time, it should not completely ignore the local differences because every region has its own culture, politics, and priorities. A balanced framework would give vendors and utilities a clear guideline—which encryption standards to use, what kind of data can be collected, how long it can be stored, and what measures should be taken to earn consumer trust (Figure 7.7).

But building something like this is not easy at all. It requires cooperation between many players—governments, regulators, power companies, equipment vendors, and even consumer groups. All of them have their own interests, and sometimes those interests clash. For example, a government may want strict monitoring for national security, while consumers may demand more privacy. Vendors may want flexible standards to reduce cost, while regulators push for stricter rules. Aligning all these different voices into one global framework is a huge task [34]–[36].

Still, if this coordination does not happen, smart grids will remain fragmented. One country's system may not fully connect with another's, and security holes may appear in the gaps. Attackers can take advantage of these weak links. That is why future research should not only look at technical solutions like encryption or blockchain, but also at policy models and legal harmonization. Incentive mechanisms may also be needed so that all players find it beneficial to follow the global framework [37]. In short, a global privacy-preserving compliance framework is not an ideal thought; it is a necessity for the next generation of smart grids. Without it, the illusion of an interconnected, secure global energy system will never be complete.

7.7 CONCLUSION

7.7.1 Recap of cyber threats in smart grids

Smart grids have evolved to be the lifeline of modern electric power system. They can add intelligence, automation, and efficiency to power distribution and consumption. But with all these benefits, smart grids also come with serious risks. In addition, with the heavy dependence on digital communication, sensors, and data collection, they are more vulnerable to cyber threats than the old traditional grids. In the past decade, researchers and governments have seen a significant increase in cyberattacks hitting critical energy infrastructure. Some of these attacks were relatively minor nuisance events, but others caused blackouts and huge financial losses.

The most common cyberattacks are the FDI attacks, in which an attacker alters the data of smart meters or sensors. This form of attack can cause the control center to think incorrect things, which can cause inappropriate decisions, such as overloading or shutting down parts of the grid. Another major threat is DoS attacks, in which communication channels are overwhelmed by extraneous traffic. Such attacks slow or stop real-time data flows, which are essential to grid stability. If the signal of the protective relay takes even a few seconds to trip, it can result in massive damage.

Smart grids are also susceptible to malware and ransomware. Attackers can infect devices or control systems, lock out users, and request ransom. Unlike normal IT, the impact here can directly affect the electricity supply, so the effects can be even more deadly. Another fast-growing threat is privacy leakage. Since smart meters collect very fine-grained consumption data, attackers can analyze this information to guess the personal habits of households. This may seem less dangerous than a blackout, but it is a huge violation of consumer trust and privacy. Attackers may not need to target the central control center directly; sometimes, attacking one small weak IoT device is enough to create a chain reaction that impacts the whole network.

In general, cyber threats to smart grids are manifold—data manipulation and DoS, Health Insurance Portability and Accountability Act (HIPAA) breach, malware intrusions, and so on. Making the issue even more difficult is that these threats keep changing. Attackers are also getting more intelligent, applying sophisticated techniques such as AI-based attacks and even training in the era of quantum computing. This implies that the traditional IT solutions cannot be used alone to secure smart grids. It needs new methods that address the special considerations of power systems with new mitigation techniques.

7.7.2 Emphasis on criticality of the five key problems

In the whole discussion in this chapter, it becomes evident that there are five main issues that are most critical concerning smart grid security and privacy. They are scalability of blockchain solutions, susceptibility of AI-based systems, lightweight cryptography constraints to resource-constrained devices, difficulties in federated learning, and decentralization of regulatory harmony. All these issues relate to a separate tier of the smart grid and collectively they define the overall risk environment. The initial problematic area is the scalability of blockchain-based solutions. Though blockchain offers permanence and reliability, it does not perform when the transaction count becomes excessively high. Millions of transactions a day will be in smart grids at the national level. When they cannot be served by the blockchain speedily, delays will be experienced, which is unacceptable in a real-time system. It makes scalability not only a technical issue but also a problem of survival, for the blockchain implementation in smart grids. The second key issue is that AI models are vulnerable. Detecting anomalies and intrusion is being carried out using AI and ML, but those models can also be tricked by an adversarial attack, or poisoned data. Assuming that the attackers can fool the AI into dismissing a genuine intrusion, the whole defense layer will fail. This issue is decisive, as an increasing amount of trust is being put on AI-based security, and in the case of failure, this can provide an opportunity for massive organized attacks.

The third problem is edge device lightweight cryptography. IoT devices and smart meters do not possess the processing capabilities of big servers. When we use heavy algorithms such as RSA, they will be too slow. In the event we apply weaker encryption, security is compromised. It is necessary to find the appropriate compromise between lightweight cryptography. Small devices will be easy to attack without it.

Challenges are the fourth problem, which is federated learning. The idea of federated learning was viewed as a way of sharing smarts without sharing data. However, it is in reality experiencing bandwidth issues, synchronization, and even information leakage due to model updates. Unless these problems are resolved, federated learning can continue to be just a hypothetical defense.

Finally, the absence of regulatory convergence in the world is a colossal obstacle. Each country possesses its own legislation, and unless these are harmonized, cross-border grids and energy trading will remain at stalemate. Interjurisdictional gaps can be exploited by attackers. It is not a legal issue but a security threat.

In taking a glance at these five problems, we can easily tell that they are interrelated. When an area is weak, then the other can be compounded. This is why their resolution is not a choice; it is the key to the living and development of safe smart grids.

7.7.3 Final note: balancing efficiency, security, and privacy is the way forward

Smart grids are like walking on a thin rope; each step demands balancing three important objectives: efficiency, security, and privacy.

Power flow is required to be instantaneous on the efficiency side, and operations are required to be speedy and smooth. Security-wise, the attackers might bring about blackouts, break equipment, or injure society in case security is lax. Privacy is also paramount; in case consumers believe that their personal habits are revealed with the help of smart-meter data, they can be resistant to implementation. These three goals do not balance readily. When our security concerns are excessive, systems may become heavy and slow. When we only focus on efficiency, weak points are created, which are vulnerable to attacks by attackers. Disregard of privacy destroys confidence. Consequently, engineers and researchers have to take into account all the three objectives simultaneously. A layered defence approach is a realistic approach. Low-level operations such as protective relays and other critical operations should be simple and fast, while higher-level operations such as billing and energy trading can apply more robust cryptography and privacy controls. The other approach is the adaptive systems. Authentication can be relatively light in ordinary use and more difficult if suspicious activity is found. Such flexibility is not the cost of efficiency or a security trade-off. The balance is further impacted by policies and regulations. Governments and utilities must adopt privacy-preserving policies, but not be so excessive that they stifle innovation. There needs to be a balance, and both consumers and providers need to be safe. In the end, the stability of the smart grids depends on this. Technology and threats are going to keep evolving. The smart grid is not only a backbone of the modern energy system, but also a means of ensuring safe, secure, and energy-efficient society.

REFERENCES

[1] Li, Z., et al. (2023, April). Blockchain-empowered security and privacy technologies for smart grids. *Computer Standards & Interfaces*, 85(C). https://doi.org/10.1016/j.csi.2022.103708

[2] Rahman, A., et al. (2025, April). An efficient blockchain-based privacy preservation scheme for smart grids. *Frontiers in Communications and Networks*, 6, 1584152. https://doi.org/10.3389/frcmn.2025.1584152

[3] Darzi, S., Akhbari, B., & Khodaiemehr, H. (2024). Improving privacy-preserving techniques for smart grid using lattice-based cryptography [Preprint]. *arXiv*. https://doi.org/10.48550/arXiv.2404.16865

[4] Zhang, Z., Rath, S., Xu, J., & Xiao, T. (2024). Federated learning for smart grid: A survey on applications and potential vulnerabilities. *arXiv preprint* arXiv:2409.10764. https://doi.org/10.48550/arXiv.2409.10764

[5] Waseem, M., Adnan Khan, M., Goudarzi, A., Fahad, S., Sajjad, I. A., & Siano, P. (2023). Incorporation of blockchain technology for different smart grid applications: Architecture, prospects, and challenges. *Energies*, 16(2), Article 820. https://doi.org/10.3390/en16020820

[6] Zhao, M., Ding, Y., Wang, B., Tang, S., Liang, H., & Wang, H. (2022). A blockchain-based framework for privacy-preserving and verifiable billing

in the smart grid. *Peer-to-Peer Networking and Applications,* 16(1), 142–155. https://doi.org/10.1007/s12083-022-01379-4.

[7] Gunathilaka, K., et al. (2025). Cybersecurity challenges in smart grids: An overview. *Energies*, 18(1), 141–152.

[8] Jha, R. K. (2023). Cybersecurity and confidentiality in smart grid for enhancing sustainability and reliability. *Recent Research Reviews Journal*, 2(2), 215–241, doi: 10.36548/rrrj.2023.2.001.

[9] Layode, A. (2024, September). Review of the literature on smart grid cybersecurity. ResearchGate Preprint.

[10] Gupta, A., et al. (2023, June). Recent advancements on cyber security for smart grids: A survey. APSIT Conference Proceedings.

[11] Alomari, M. A., Al-Andoli, M. N., Ghaleb, M., Thabit, R., Alkawsi, G., Alsayaydeh, J. A. J., & Gaid, A. S. A. (2025). Security of smart grid: Cybersecurity issues, potential cyberattacks, major incidents, and future directions. *Energies,* 18, 141. https://doi.org/10.3390/en18010141

[12] TechRadar. (2025). Millions of solar power systems are at risk of cyber attacks. *TechRadar.*

[13] Zhang, Y., Wang, J., & Liu, H. (2023). Lightweight cryptographic techniques for secure smart meter communications. *IEEE Transactions on Smart Grid,* 14(3), 1992–2003.

[14] Fernandez, D., & Miller, T. (2023). Legal harmonization for international energy data exchange. *Journal of Cyber Policy*, 8(2), 231–248.

[15] Kumar, S., & Singh, P. (2024). AI-driven intrusion detection systems for smart grid applications. *Journal of Information Security and Applications*, 78, 103515.

[16] Zhao, L., Chen, R., & Wu, K. (2023). Federated learning in energy systems: Security challenges and opportunities. *IEEE Access,* 11, 98576–98590.

[17] Kumar, G., Manoharan, G., Arora, K., and Kumar, S. Optimizing Heart Disease Diagnosis with K-Nearest Neighbors and Logistic Regression Models. In *2025 International Conference on Sustainability, Innovation & Technology (ICSIT)*, Nagpur, India, 2025, pp. 1–4, doi: 10.1109/ICSIT65336.2025.11294937.

[18] Das, S., et al. (2023). Adversarial machine learning threats to smart grid security. *IEEE Internet of Things Journal*, 10(5), 4021–4032.

[19] Oliveira, M., & Costa, J. (2023). Privacy leakage risks in smart meter data: A systematic review. *Energy Reports*, 9, 1201–1215.

[20] Hou, D., Zhang, J., Huang, S., Peng, Z., Ma, J., & Zhu, X. (2022). Privacy-preserving energy trading using blockchain and zero knowledge proof. In *Proc. 2022 IEEE International Conference on Blockchain (Blockchain)*, pp. 412–418.

[21] European Commission. (2023). GDPR and smart energy data protection guidelines. Official Report.

[22] North American Electric Reliability Corporation (NERC). (2024). Critical infrastructure protection (CIP) standards. NERC Technical Report.

[23] Wang, P. (2023). Quantum threats to energy infrastructure: Preparing smart grids for the post-quantum era. *Energies*, 16(14), 5315–5332.

[24] Patel, A., & Roy, D. (2024). Lightweight lattice-based encryption for IoT-enabled smart grids. *ACM Transactions on Cyber-Physical Systems*, 8(1), 1–22.

[25] Mehta, R. (2023). Hybrid blockchain architectures for energy trading platforms. *Journal of Grid Computing*, 21(2), 335–350.

[26] Ahmed, K. (2024). Scalability issues in blockchain-enabled smart grids: A survey. *IEEE Access*, 12, 67412–67428.

[27] Liu, J., & Zhang, F. (2023). Hardware-assisted security for smart meters using secure elements. *Computers & Electrical Engineering*, 110, 108774.

[28] Sharma, V., & Kaur, M. (2024). Denial of service attacks in smart grids: Detection and mitigation techniques. *International Journal of Electrical Power & Energy Systems*, 152, 109309.

[29] Choi, J., et al. (2023). Resilient edge computing for smart grid security. *IEEE Transactions on Industrial Informatics*, 19(7), 7652–7661.

[30] Singh, P., Arora, K., & Rathore, U. C. (2022). Control strategies for improvement of power quality in grid connected variable speed WECS with DFIG–An overview. *Journal of Physics: Conference Series, 2327*(1), 012008. https://doi.org/10.1088/1742-6596/2327/1/012008.

[31] Lee, S., & Park, M. (2024). Adaptive authentication models for smart grid cybersecurity. *Journal of Network and Computer Applications*, 227, 103635.

[32] Channi, H. K., Sandhu, R., & Anand, A. (2026). Integration of Smart Grids in Energy Optimization. *Optimization in Sustainable Energy: Methods and Applications,* 329–356. https://doi.org/10.1002/9781394242139.ch10

[33] Verma, A. (2024). AI-based adaptive privacy control in IoT-driven smart grids. *Future Generation Computer Systems*, 154, 186–198.

[34] Bhatt, B., Sharma, H., Arora, K., Joshi, G. P., & Shrestha, B. (2023). Levy flight-based improved grey wolf optimization: A solution for various engineering problems. *Mathematics, 11*(7), 1745. https://doi.org/10.3390/math11071745.

[35] Jensen, T. (2023). Consumer trust and data ownership in smart energy systems. *Renewable and Sustainable Energy Reviews*, 178, 113227.

[36] Channi, H. K., Handa, K., Bhullar, M., Sandhu, R., Kaur, H., and Singh, P. Renewable Distributed Generation for Smart Highway Lighting. In *2024 11th International Conference on Reliability, Infocom Technologies and Optimization (Trends and Future Directions) (ICRITO)*, pp. 1–6. IEEE, 2024.

[37] United Nations Energy Council. (2025). Towards a global framework for privacy-preserving smart grid interoperability. Policy Paper.

Chapter 8

Energy storage systems and grid flexibility

Pradeep Singh and Krishan Arora

8.1 INTRODUCTION

8.1.1 Background and motivation

There is a paradigm shift in the energy landscape of the world with the twin aims of decarbonization and sustainability. Countries are working hard to minimize emissions of greenhouse gases and achieve international climate requirements, and this has greatly led to the use of renewable energy sources like solar photovoltaic (PV), wind, and hydro. Such sources are inherently non-polluting, and most of them are available in large quantities, and their nature of variability and intermittency creates distinct challenges to the stability, reliability, and operational flexibility of the electrical power systems [1]. The conventional power generation was based on predictable, dispatchable generation by fossil fuel plants. These systems had a one-way flow of electricity, generation at one place, and passive consumers. The development of the distributed energy resources (DERs), especially renewables, has, however, presented the possibility of two-way energy flows and more dynamic, adaptive system operation. The flexibility of grid, the capacity of power system to respond quickly and effectively to supply and demand changes has become very important in this new paradigm. A major facilitator of this flexibility has been found in the energy storage systems (ESS). They provide the advantage of decoupling generation and consumption through the storage of excessive energy when the demand is low or when the generation is high and when one is required. This capability deems ESS invaluable to balancing supply and demand, as well as to increasing grid reliability, frequency, and voltage regulation support, renewable integration, and costly grid infrastructure deferral.

8.1.2 What is grid flexibility?

The term grid flexibility is used to refer to the ability of the power system to handle the anticipated and unforeseen shifts in supply and demand

DOI: 10.1201/9781003689430-8

without incurring critical and unreliable operations. Different resources can provide flexibility, including traditional generators, responsive loads, interconnectors and, more and more, energy storage. It is necessary on a variety of timescales: within the range of a few milliseconds (e.g., stabilization of frequency) to hours (e.g., maxima shaving every day) and also seasons (e.g., storing up excess solar energy during the summer months to use in winter). This growing complexity of electricity networks that have large proportions of variable renewable energy (VRE) requires solutions capable of operating over these time scales [2]. The variety of energy storage and its scalable character make it the uniqueness of storage to meet these functions. An example is that lithium-ion batteries are good at short-duration frequency support, and pumped hydro and hydrogen storage can offer long-duration energy-shifting capacity.

8.1.3 The role of energy storage in modern power systems

Implementation of ESS is not a dream anymore- it is a reality. Contemporary ESS technology is currently being implemented on residential and commercial, and utility scales. They supply a large number of grid functions, such as:

- Discharge during peak-demand periods.
- Firming of renewable energy through reduction of generation variability.
- Black start ability through system restoration without an external power source.
- Transmission and distribution (T&D) support through decongestion and delaying upgrades.
- Energy arbitrage using energy storage to store it when prices are low and set it when prices are high.

Additionally, due to the development of smart grids and the digital era, ESS is becoming more and more a part of intelligent energy management systems, virtual power plants (VPPs), and demand response programs, which makes electricity networks increasingly responsive and efficient [3].

8.1.4 Structure of the chapter

This chapter explores the essence of energy storage and its importance in increasing grid flexibility. It starts with categorizing the different kinds of ESS on the basis of physical and chemical mechanics of operations. The following details performance measures and specific implementation of power systems. Particular attention is paid to the role that ESS play in grid flexibility and the renewable integration [4]. The actual practice, including real-life cases and trends of deployment, offers some practical information about the current activities in the world. Lastly, the subject of the chapter touches on technical, regulatory, and economic issues, whereby future prospects and innovations in energy storage technology are also discussed.

8.2 CLASSIFICATION OF ESS

ESS are categorized according to the energy storage form and the physics or chemistry behind the process. The different types of storage systems possess different features that are applicable in various power grid applications—frequency control down to milliseconds and energy shifting over a season. In this section, the researcher has given a summary of the key forms of energy storage technologies.

8.2.1 Mechanical ESS

The storage systems in which the energy is stored in physical form are mechanical and normally by altering the position or movements of mass. These systems are older and have one of the most established technologies.

8.2.1.1 Pumped hydro storage (PHS)

PHS is the largest-scale storage that is globally used. It operates by pumping the water from a low reservoir and into a high reservoir when the demand for electricity is low. In periods of high demand, water will be released to the upper reservoir through turbines to produce electricity [5].

- Merits: High efficiency (7085%), long service life, large power (GW scale).
- Limitations: Site-dependent, expensive to construct, construction time-consuming.

8.2.1.2 Compressed air energy storage (CAES)

Air is pressurized and stored in underground caverns or tanks in CAES systems. The compressed air is then heated and allowed to expand in a turbine to produce electricity when energy is required.

- Pros: It is suitable for storing in bulk, scalable. Limitations: Needs special geological formations, less efficient (40–60%)

8.2.1.3 Flywheel energy storage (FES)

Flywheels store energy in the form of kinetic energy as rotation of a mass at high speed. When charging, the electricity is transformed to kinetic energy, and when discharging, it is transformed back to electricity through slowing of the rotor. The benefits include: High power density, exceptionally short response times, cycle life is long.

- Disadvantages: Low power potential, high self-discharge.

8.2.2 Electrochemical ESS

Such systems accumulate energy in a chemical form and discharge it by means of electrochemical reactions. These are traditional batteries and new innovative technologies of batteries.

8.2.2.1 Lead-acid batteries

A fully grown and economical technology that is frequently applied in backup power and uninterruptible power supplies (UPS).

- Advantages: High surge currents, low cost.
- Disadvantages: Low energy density, short cycle life, environmental issues with lead.

8.2.2.2 Lithium-ion batteries

The most popular battery technology in electric cars and stationary storage today. It provides high power and energy density such that it can be used on a grid scale.

- Pros: 90–95% efficiency, long life cycle, small size.
- Limitations: Thermal runaway: Thermal runaway is a risk, high cost (and falling): Lithium, cobalt, resource constraints

8.2.2.3 Flow batteries

Flow batteries (e.g., vanadium redox flow) have liquid electrolytes in external tanks, and these are pumped through electrochemical cells, and in operation.

- Benefits: Scalability (decoupled with power) of the energy capacity, long cycle life.
- Limitations: low energy density, complicated systems.

8.2.3 Electrical ESS

These systems simply store energy in electrical or magnetic fields and have an extremely fast response time and power densities.

8.2.3.1 *Supercapacitors (ultracapacitors)*

More electric double layers Store energy by the separation of electric charges. Useful in applications where there is a need to charge/discharge fast.

- Benefits: high power density, short cycle time, long life cycle.
- Limitations: Low energy density, self-discharge.

8.2.3.2 *Superconducting magnetic energy storage (SMES)*

Stores energy in the magnetic field within a superconducting coil as a result of direct current passing through the coil. Is able to emit energy near instantly.

- Benefits: Very high speed of response, efficiency.
- Disadvantages: Pricey, must be cooled with cryogenics, low power capacity.

8.2.4 Thermal ESS

Thermal storage systems conserve energy in the form of heat and discharge it when it is required [6]. They can be applied in power and heating/cooling.

8.2.4.1 *Sensible heat storage*

Stored energy is the increase of temperature of a solid or liquid substance (e.g., molten salts, water, rocks). Examples: Concentrated solar power (CdS) plants have molten salt tanks.

- Pros: cheap and easy to design.
- Limitations: Limit on material heat capacity and temperature range.

8.2.4.2 *Latent heat storage*

Employs phase change materials (PCMs) that store/release energy in the process of melting or solidification.

- Benefits: high energy density, isothermal behavior.
- Limitations: Material degradation, increased cost.

8.2.4.3 *Thermochemical storage*

Reversible chemical reactions to store/release heat energy. In the experimental phases but has high energy density and potential life duration.

8.2.5 Chemical ESS

The conversion of electricity into chemical fuels, including hydrogen, ammonia, or synthetic hydrocarbons, can be stored and reused in other areas (mobility, industry).

8.2.5.1 Hydrogen storage (power-to-gas)

Water is electrolyzed using electricity to get hydrogen that can be stored and consumed later in the fuel cell or combustion turbine.

- Benefits: Large power potential, cross-sectoral application (power, transport, industry).
- Weaknesses: Round-trip low efficiency (3040%), infrastructure and safety issues.

8.2.5.2 Synthetic fuels (e-fuels, ammonia)

Replacing carbon or nitrogen with hydrogen to produce synthetic hydrocarbons or ammonia, which have the potential to be carbon-neutral fuels.

- Advantages: The ability to use the same fuel infrastructure.
- Disadvantages: Low efficiency, high energy input (Table 8.1).

This typology forms the basis of the knowledge of the most suitable storage technologies for certain grid applications, time scale, and business goals.

Table 8.1 Summary table: energy storage technologies at a glance

Technology	*Energy form*	*Efficiency (%)*	*Duration*	*Applications*
Pumped Hydro	Mechanical	70–85	Hours to days	Bulk storage, peak shaving
Li-Ion Battery	Electrochemical	90–95	Minutes to hours	Grid Balancing, EVs
Flow Battery	Electrochemical	60–80	Hours	Renewable Integration
Supercapacitor	Electrical	90–98	Seconds	Frequency Regulation
Thermal (Molten Salt)	Thermal	70–80	Hours to days	CSP Plants
Hydrogen	Chemical	30–40	Weeks to months	Seasonal storage, mobility

8.3 OPERATING PRINCIPLE AND PERFORMANCE METRICS

8.3.1 Operating principles of energy storage technologies

All the ESS operate on different physical, chemical, or electrical principles. The common modes of operation are listed below:

8.3.1.1 Mechanical systems

PHS works by transforming electrical energy into potential energy by pumping the water to an elevated level [7]. When it is required, the water that is kept is discharged to rotate turbines that produce electricity.

- **CAES:** CAES is used to store energy by compressing air into underground reservoirs when not in peak demand. The air is released, warmed (typically by burning natural gas), and expanded to create electricity, during peak demand.
- **Flywheels:** Flywheels store energy kinetically as a result of accelerating a rotor to high velocities. Slowing of the rotor retrieves energy by turning a generator.

8.3.1.2 Electrochemical systems

A battery (i.e., lithium-ion or lead-acid) holds energy in a reversible electrochemical reaction between an electrolyte and electrodes [8]. Charging makes the ions move and store energy; discharging makes the energy release in an organized reaction.

8.3.1.3 Electrical systems

Supercapacitors save energy as a result of electrostatic charge separation. They are able to charge and discharge extremely fast yet store a relatively small amount of energy. SMES: Superconducting magnets store energy within the magnetic field generated by a DC passing through a superconducting coil, and cooled in a cryogenic system.

8.3.1.4 Thermal systems

Sensible and latent heat systems are ESS that store energy through the elevation of temperature or transformation of phase of a material (e.g., a solid becomes a liquid). Also known as thermochemical systems, they store energy in the form of endothermic and exothermic reactions that are made and broken through the breaking and formation of chemical bonds [9].

8.3.1.5 Hydrogen systems

These systems break water down into hydrogen and oxygen through electrolysis. The hydrogen is then able to be stored, transported, and refueled in fuel cells or turbines to recreate electricity.

8.3.2 Key performance metrics

The choice of the proper storage system involves an assessment of the different performance indicators. The most important metrics are listed below:

8.3.2.1 Round-trip efficiency (RTE)

- Measures the efficiency between energy output and the energy input during one cycle of a charge–discharge process.

$$RTE = \frac{Energy\ Out}{Energy\ In} x\ 100\%$$

- Lithium-ion batteries have typical values of 90–95, pumped hydro of 70–85, and hydrogen of approximately 30–40 %.

8.3.2.2 Power rating and energy capacity

Power rating (kW or MW): maximum power the system is capable of producing at any given moment.

- Energy capacity (kWh or MWh): amount of energy that the system can hold.
- Duration (hours): Time that the storage system can maintain its rated power output.
- Short-duration: a few minutes to seconds (e.g., supercapacitors).
- Medium-duration: 14 hours (e.g., Li-ion).
- Long-term: 6 hours to days or weeks (e.g., PHS, hydrogen).

8.3.2.3 Energy density and power density

- **Energy density**: Energy stored per unit volume or mass (Wh/kg or Wh/L). High energy density is essential for space-limited applications like electric vehicles (EVs).
- **Power density**: Power delivered per unit volume or mass (W/kg). Important for high-demand applications like frequency regulation.

8.3.2.4 *Cycle life and calendar life*

- **Cycle life:** Number of charge/discharge cycles that the system can complete before the capacity decreases to a predetermined value (typically 80% of original).
- **Calendar life:** Life of a storage system regardless of the use normally due to aging, corrosion, or wear.
- Examples: Li-ion: 3,000–10,000 cycles, flow batteries: 10,000+ cycles, lead-acid: 500–1,000 cycles.

8.3.2.5 *Depth of discharge (DoD)*

- Percent of power drawn out of the system against its total capacity.
- Impairs the lifespan of the battery- deep discharges lower the life.
- DoD is often constrained by optimum operation (e.g., 80% of lithium-ion).

8.3.2.6 *Response time*

- The time spent going through the standby to full discharge.
- Fast response (less than 1 second) is an important factor of ancillary services such as frequency regulation.
- SMES and supercapacitors are almost instant in reaction.

8.3.2.7 *State of charge (SoC)*

- The present power output in relation to full power.
- As with a fuel gauge of storage systems.
- Precise SoC estimation is crucial to quality operation and battery management system (BMS).

8.3.2.8 *Self-discharge rate*

The proportion of energy that is wasted daily/weekly when in an idle position. Batteries and capacitors are not all self-discharging in the same way:

- Li-ion: ~2–3% per month.
- Supercapacitors: far higher (20–10% per day).
- PHS and CAES: negligible.

8.3.2.9 *Cost metrics*

- Capital cost (CAPEX): The first installation cost of the system (equipment, building, etc.).

Table 8.2 Ideal technology as per requirement

Requirement	*Ideal technology*
Rapid frequency regulation	Supercapacitors, SMES
Daily peak shaving	Lithium-ion batteries
Renewable energy firming	Flow batteries, Li-ion
Long-duration storage	Pumped hydro, Hydrogen
Mobile applications (EVs)	Lithium-ion batteries
Low-cost seasonal storage	Hydrogen, CAES

- Operating cost (OPEX): cost of running and maintaining the system throughout the life.
- Levelized cost of storage (LCOS): Cumulative overall cost of energy delivered on a per-unit basis.

$$LCOS = \frac{Total\ Lifetime\ Cost}{Total\ Energy\ Discharged\ over\ Lifetime}$$

8.3.3 Trade-offs and selection criteria

No single optimal technology of storage. All of them are good and bad depending on their use (Table 8.2).

The best system should be cost-effective, efficient, have a long life, and possess the technical features needed in the chosen application.

8.4 APPLICATIONS OF ENERGY STORAGE IN POWER SYSTEMS

ESS are becoming very critical to the contemporary power grid. Capacity to store electricity in plentiful times and discharge it in periods of scarcity makes them a flexible resource to system operators, utilities, prosumers, and factory users. ESS have extensive services in the fields of generation, transmission, distribution, and consumption [10]. The section will divide the major uses of ESS in current power systems into four broad categories: grid services, renewable integration, load management, and infrastructure optimization.

8.4.1 Grid services

The importance of energy storage to grid stability is that it supplies ancillary services, which are essential in ensuring safe and effective operations of the power system.

8.4.1.1 *Frequency eegulation*

- ESS is able to inject or absorb power quickly so that the grid frequency can stay within allowable limits (e.g., 50 Hz in India, 60 Hz in the United States).
- The response of the traditional generators to the deviations of the frequency is slower than the responses of the batteries (in particular, lithium-ion): the batteries can start reacting within milliseconds.
- Examples: Pennsylvania–New Jersey–Maryland Interconnection (PJM) (U.S.) has a frequency regulation market in which battery storage generates revenue through frequency balancing.

8.4.1.2 *Voltage support*

- ESS has the potential to offer reactive power to stabilise grid voltage levels.
- Voltage problems usually happen in poor distribution systems or when there is a lot of renewable energy generation.
- Storage with inverters can be used as static synchronous compensators (STATCOMs).

8.4.1.3 *Black start capability*

- After blackout occurs in the entire system, ESS can be used to jump-start the power system with substations and auxiliary systems energized [11].
- As opposed to traditional black start units, batteries do not need fuel or spinning reserve to boot.

8.4.1.4 *Spinning and non-spinning reserve*

- ESS is capable of generating a quick-ramping backup power generation competent to substitute the traditional generators.
- Their fast-time of starting makes them the best in case of contingency reserves.

8.4.2 Renewable energy integration

With the introduction of VRE sources like wind and solar energy onto the grid, storage becomes an important part of reducing their intermittency and enabling seamless integration [12].

8.4.2.1 *Smoothing and firming of renewable output*

- Renewable generation is day-neutral and varies at different times of the day.

- Short-term variation can be evened out by ESS and output made more predictable and dispatchable.

8.4.2.2 *Energy shifting*

- ESS stores surplus generation when the production is too great (e.g., midday solar) and releases it when demand is the highest (e.g., evening hours).
- This increases the worth of renewables as it balances supply and demand.

8.4.2.3 *Curtailment reduction*

- In high-renewable systems, excess generation is frequently curtailed (unaffiliated) because of network capacity.
- Storage can absorb this extra energy, minimize waste, and increase economical returns.

8.4.2.4 *Grid stability with high VRE penetration*

- High solar and wind levels cause a decrease in system inertia, thus causing the grid to be susceptible to instability.
- Advanced inverter ESS is capable of synthetic inertia and rapid frequency response.

8.4.3 Load management and demand optimization

Energy storage plays a significant role in the administration of electricity demand, peak loads, and energy efficiency.

8.4.3.1 *Peak shaving*

- ESS releases when there is peak demand so as to decrease load on the grid or to avoid high energy charges.
- This is particularly useful to commercial/ industrial consumers who are subject to time-of-use (TOU) tariffs or demand charges [13].

8.4.3.2 *Load leveling*

- It entails charging the battery when the cost is cheaper and discharging when it becomes expensive.
- It makes the demand curve flat, increasing the efficiency of the system on the whole.

8.4.3.3 Behind-the-meter (BTM) storage

- ESSs at the customer location (homes and businesses): ESSs offer grid independence, self-consumption optimization, and backup power.
- Combined with rooftop solar, it allows an increase in self-reliance and resilience.

8.4.3.4 Demand response

- ESS may be involved in demand response programs by lowering net demand at peak hours or supplying stored power to the grid.
- Small-scale systems can be made to appear as a large dispatchable unit through the use of aggregated storage (e.g., via VPPs).

8.4.4 T&D optimization

Storage is able to alleviate strain on the T&D system, improving grid reliability and delaying expensive investments.

8.4.4.1 Congestion relief

- Transmission lines might be congested during high times.
- Storage where grids are strategically located may be discharged to satisfy local demand and decrease the flow of power along restricted corridors.

8.4.4.2 Infrastructure deferral

- And ESS can delay transformer, substation, and feeder upgrades by decreasing peak demand.
- This results in savings to utilities in terms of capital expenditure (CAPEX).

8.4.4.3 Voltage and power quality management

- Storage is used in low-voltage or rural networks to stabilize the voltage and to reduce power quality problems in changeable loads or renewables.

8.4.4.4 Islanding and microgrids

- ESS makes microgrids grid-forming, which means that they can function independently (islanded mode) when outages happen.
- Storage-backed microgrids enhance access to energy and resilience in remote or disaster pragmatic regions (Table 8.3).

Table 8.3 Illustrative examples

Application area	*Example technology*	*Typical use case*
Frequency regulation	Li-ion, flywheels	Real-time balancing of generation and load
Renewable energy integration	Li-ion, flow batteries	Shifting solar output to evening demand
Load management	Residential batteries	Reducing peak consumption and TOU charge optimization
Grid investment deferral	Utility-scale batteries	Avoiding substation upgrades in growing urban areas
Microgrid support	Hydro solar + battery	Ensuring uninterrupted power in remote villages

8.5 GRID FLEXIBILITY AND ITS DIMENSIONS

8.5.1 Introduction to grid flexibility

With the accelerated shift in the modern power systems brought on by renewable power, electrification, and decentralization, grid flexibility is a key necessity to ensure reliability and economic efficiency. The classical grids were modelled based on predictable and dispatchable generation based on fossil fuels. By contrast, the current grids need to accommodate the very changeable and uncertain contributions of solar and wind sources, as well as other renewable sources. This renders the flexibility or the capacity of the grid to react to the supply and demand changes in different time frames and locations a vital operational feature. With a flexible grid, system operators are in a good position to effectively absorb uncertainty, variability, and unexpected fluctuations in electricity demand or supply [14]. With flexibility, there is less need to curtail, no blackouts, and a safe energy move over.

8.5.2 Definitions and key concepts

The concept of grid flexibility is the capability of a power system to change the generation, demand, or storage, effectively responding to the net load (load, less variable renewable generation), without upsetting system balance and reliability. One can describe it in three main directions:

- **Temporal flexibility:** The capacity to harmonise energy across time, seconds, minutes, hours, days, and seasons.
- **Spatial flexibility:** This refers to the capability of movement of energy to various geographic locations through the T&D system.
- **Resource flexibility:** Flexibility to utilize a combination of flexible resources such as generation, demand-side management, and storage.

8.5.3 Sources of grid flexibility

Current-day power systems are based on several technologies and approaches to reaching the necessary degree of flexibility:

8.5.3.1 Flexible generation

Conventional gas turbines and hydro facilities can usually follow fluctuations in demand by quickly switching through on and off. Older generators with slow start/stop are being displaced by newer ones, which are designed to start and stop quickly.

8.5.3.2 Responsive demand (demand-side flexibility)

Consumers decrease or shift consumption of electricity according to the price signals or incentives. Demand response programs- can call industrial or residential loads into effect during peak demand or system stress.

8.5.3.3 Energy storage systems

ESS is able to store surplus electricity in times of low demand and discharge during times of high demand.

8.5.3.4 Storage

Storage provides flexibility in the short-term and in the long-term (e.g., batteries versus hydrogen storage).

8.5.3.5 Grid infrastructure (flexibility in the T&D)

The better the grid interconnections and smart grid technologies improve the space management of the electricity. Highly distributed automation and reconfigurations of the network enhance local reliability and flexibility.

8.5.3.6 Market mechanisms

The flexibility of all actors is encouraged by real-time pricing, ancillary service markets, and capacity markets. TOU tariffs, net metering, and aggregation of DERs are applied in optimization of demand and supply behavior.

8.5.4 Dimensions and time scales of flexibility

Refer Table 8.4 for dimensions and time scales flexibility.

Table 8.4 Details of timeframe for various flexibility need

Timeframe	*Flexibility need*	*Example of technologies*
Seconds	Frequency stabilization	Supercapacitors, battery inverters, SMES
Minutes	Regulation reserves	Lithium-ion batteries, fast-ramping generators
Hours	Load shifting, peak shaving	Batteries, pumped hydro, flow batteries
Days to weeks	Managing low renewable output	CAES, hydrogen storage, thermal storage
Seasonal	Matching supply-demand over months	Hydrogen, Power-to-gas, large hydro reservoirs

8.5.5 Indicators to quantify flexibility

The measurement of grid flexibility assists in the planning and operation of energy systems. Common metrics include:

- **Ramp rate (MW/min):** Capacity of generators or storage systems to add or reduce output.
- **Reserve margin (%):** A surplus capacity to cater to sudden demand.
- **Factor of flexibility:** Fraction of system demand that is satisfied by flexible resources. The demand curve varies after considering the renewable generation: this is called the net load variability.
- **Curtailment rate:** Renewable energy is not utilized because of the unavailability of flexibility.

8.5.6 Role of digitalization and smart grids

Digital technologies, including advanced metering infrastructure (AMI), real-time data analytics, artificial intelligence (AI)-based control, and Internet of Things (IoT)-based automation, have made grid resources more visible, forecastable, and controllable. These innovations would be necessary to achieve latent flexibility in the current assets and improve coordination in the grid.

8.6 ENERGY STORAGE AS A FLEXIBILITY ENABLER

8.6.1 The central role of storage in flexible grids

One of the most potent mechanisms that could be used to increase the grid flexibility is ESS. Compared to conventional generation or demand response systems, storage is adaptable to a variety of time scales- meaning milliseconds to months- and is therefore unique. It is able to take, store, and inject power when needed and is both a load and a generator. Storage is useful to balance the sometimes-fickle relationship between supply and demand as grids begin to shift away towards more intermittent renewable energy sources, stabilizing frequency and voltage, and offering capacity and contingency reserves. Stored co-located with generation, with the grid, or installed BTM, storage is making flexibility more flexible than it ever was before, faster than before, more accurate than before, and more controllable than before.

8.6.2 Short-term vs. long-term flexibility through storage

With application and grid requirements, energy storage technologies can be implemented to provide short-duration or long-duration flexibility.

8.6.2.1 Short-term flexibility (seconds to hours)

Systems to offer instant support to grid balancing: Short-duration systems offer instantaneous support to grid balancing: Frequency regulation: Lithium-ion and supercapacitors give responses in the sub-second range. Spinning capacity: Gas turbines can be substituted with storage to serve as contingency reserves. Voltage support: Voltage control off-grid voltage inverters provide grid voltages.

8.6.2.2 Long-term flexibility (hours to weeks or seasons)

Long-duration energy storage (LDES) is used to cope with both multi-day imbalances and seasonal energy changes:

- **Pumped hydro, compressed air, and flow batteries:** 6–12+ hour discharge times.
- **Hydrogen and power-to-gas:** Store extra electricity in weeks or months.
- **Thermal energy storage:** Saves solar energy, which can be used in district heating or in future production of electricity (Table 8.5).

8.6.3 Distributed energy storage and the flexibility ecosystem

Storage of energy is no longer limited to the central utilities. Storage is being distributed at homes, businesses, and communities with the emergence of DERs.

Table 8.5 Storage technologies as per duration

Duration	*Storage technologies*	*Key use cases*
Seconds	Supercapacitors, SMES	Frequency regulation, stability
Minutes	Flywheels, Li-ion	Spinning reserve, grid balancing
Hours	Li-ion, flow batteries, pumped hydro	Peak shaving, load shifting
Days	CAES, thermal storage	Renewable firming, resilience
Weeks	Hydrogen, ammonia, synthetic fuels	Seasonal storage, backup power

8.6.3.1 BTM Storage

- Mounted in residential locations (e.g., Tesla Powerwall, Sonnen).
- Applications: used to optimize self-consumption, to establish backup power, and to reduce demand charges.
- When pooled together in VPPs can offer grid services.

8.6.3.2 Campus and community storage

- Common storage resources used by microgrids or campuses or rural electrification.
- Promotes local reliability, lessens peak demand charges, and advances energy independence.

8.6.3.3 Front-of-the-meter (FTM) storage

- Utility-run or independent grid-scale storage.
- Provides ancillary services, congestion relief, and integration of renewable at T&D levels.

8.6.3.4 Aggregated and mobile storage

- Small-scale batteries may be combined through digital means to act as one larger unit.
- EV batteries and second-life batteries could be repurposed or deployed dynamically to offer mobile storage to grid.

8.6.4 Energy storage in flexibility markets

Storing energy is becoming an active member of electricity markets, not only as a fixed part of the asset but as an active market player.

8.6.4.1 Energy arbitrage

- Purchase power when it is cheap (off-peak) and sell when it is expensive (peak).

- Needs proper forecasting, market entry, and modeling of finances.
- Examples: Battery storage fleets are cost-effective with respect to energy time-shifting in California.

8.6.4.2 Ancillary services

- Storage is a profitable venture because grid services, including frequency regulation, spinning reserves, and voltage control, are provided.
- Batteries are suitable for these services because of their quick-ramping capabilities.

8.6.4.3 Capacity markets

- ESS would have the ability to be involved in capacity auctions that ensure it is available during peak hours.
- This will push out fossil-based peak load plants and will make the systems more efficient.

8.6.4.4 Demand response and flexibility platforms

- Storage assets are being incorporated in flexibility platforms and real-time balancing markets.
- In Europe, sites such as PicloFlex and EPEX Spot have the potential to provide a chance of decentralized storage to enable demand-side flexibility.

8.6.5 Case example: VPPs

A VPP gathers thousands of small and distributed storage facilities (and occasionally solar, EVs, and flexible loads) and coordinates them with a central control platform.

- **Function:** Operates as a big dispatchable facility that is able to sell energy or services at the grid.
- Merits: Small-scale assets unlock value, decrease peak loads, and enhance grid resiliency.
- Real-world examples:
  - Next Kraftwerke: Aggregates storage, wind, and biogas plants, Germany.
  - Tesla VPP in Australia: The grid is stabilized by thousands of home batteries.

8.6.6 Grid services enabled by energy storage

Table 8.6 provides the grids services.

Table 8.6 Grid service as per storage role

Grid service	*Storage role*
Frequency regulation	Fast response from batteries and flywheels
Voltage support	Inverter-based reactive power control
Black start capability	Enabling grid restart without fuel-based generation
Peak shaving	Reducing grid stress and electricity costs
Load shifting	Balancing daily load profiles and renewable variability
Congestion relief	Discharging near demand centers to avoid transmission overload
Renewable energy firming	Smoothing solar/wind output and reducing curtailment
Backup power	Ensuring reliability during outages or disasters

8.6.7 Benefits of using storage for flexibility

- **Speed and accuracy:** Batteries react in milliseconds—perfect to be used in real-time balancing.
- **Decentralization:** It may be implemented in any location—homes, business, substations.
- **Scalability:** Between utility-scale and small residential units.
- **Multi-functionality:** There are multiple roles that a single storage system can carry out, either simultaneously or in sequence.
- **Resilience:** Assists with blackouts, grid disturbances, and emergency services.

8.7 CASE STUDIES AND GLOBAL DEPLOYMENT TRENDS

8.7.1 Introduction

In the last 10 years, storage of energy has ceased to be an experimental pilot project but a mainstream infrastructure. Nations of the world are implementing large-scale and distributed storage systems to assist in the integration of renewable energy, provide grid stability, and provide flexibility in the energy market. This part provides key case study examples in prominent areas and reviews the new trends in the world in terms of technology usage, policy development and investment policies [15].

8.7.2 United States: grid-scale storage in California

Project: *Crimson Energy Storage Project, California*

- **Capacity:** 350 MW/1,400 MWh.
- **Technology:** Lithium-ion battery.
- **Commissioned:** 2023.

Significance
California is the leader in the energy storage implementation to a great extent due to ambitious renewable energy targets and accommodating market designs. The Crimson Project is one of a larger scale of stabilizing the grid under conditions of high solar penetration, notably during the infamous duck curve effect, when daytime sun triggers high ramping requirements in the evening.

Impact

- High response grid balancing.
- Lessens solar energy wastage.
- Improves the adequacy of resources in peak demand.

Policy driver: The Resource Adequacy and Self-Generation Incentive Program (SGIP) have caused utility-scale and BTM resources to adopt storage.

8.7.3 Germany: VPP and energy transition

Project: *Next Kraftwerke VPP*

- **Components:** Over 10,000 distributed units including batteries, biogas, solar, wind.
- **Control center:** Real-time dispatch and aggregation.

Significance
The energy transition (Energiewende) of Germany focuses on distributed flexibility. Next Kraftwerke VPP demonstrates the situation when thousands of small systems can act as a giant dispatchable participant that can provide ancillary services and take part in the energy markets.

Impact

- Shows that distributed storage provides grid services.
- Lessens the use of fossil fuel peak load power plants.
- Improves the income sources of small manufacturers.

Policy driver: Liberated balancing markets and incentives to DER to be a participant in flexibility markets.

8.7.4 Australia: residential batteries and grid resilience

Project: *Tesla Virtual Power Plant (VPP), South Australia*

- **Target:** 50,000 home batteries (over 4,000 already deployed).
- **Combined Capacity:** 250 MW when complete.
- **Technology:** Tesla Powerwall + solar rooftop.

Significance
South Australia has adopted distributed energy and storage due to the problem of grid reliability and blackouts. The Tesla VPP combines residential batteries to provide grid services.

Impact

- Cuts bills and dependence on centralized generation by consumers.
- Evidences residential storage as a grid asset.
- Enhances grid resilience and blackout recovery.

Policy driver: Federal battery subsidies and South Australia's energy transition strategy.

8.7.5 India: Storage for renewable integration and rural electrification

Project 1: *Solar + Battery Microgrids (UP, Bihar)*

- Deployed by: Tata Power, Husk Power.
- Use: Rural microgrids with lithium-ion and lead-acid storage.

Impact

- Offers 24/7 power supply in off-grid areas.
- Promotes agriculture and small business.
- Cuts diesel generator reliance.

Project 2: *RE Integration Pilot—Ladakh and Himachal Pradesh.*

- Goals: Store surplus hydro/solar during evening peak times.
- Agencies: SECI, NTPC, Himurja (HP), MNRE.

Significance
Intermittent solar and wind increase the challenges in India power grid, particularly in weak-grid or remote areas. Storage is a critical factor in enhancing the quality of supply and in grid balancing.

Policy driver: National Mission on Advanced Battery Storage and Production-Linked Incentive (PLI) Scheme of battery manufacturing.

8.7.6 China: dominance in battery manufacturing and utility-scale storage

Project: *Dalian Vanadium Redox Flow Battery Project*

- **Capacity:** 100 MW / 400 MWh.
- **Technology:** Vanadium redox flow.
- **Use case:** Grid peak shifting and renewable smoothing.

Significance China not only excels in the deployment of solar PV but also manufactures and installs energy storage. Flow batteries provide a substitute for lithium in the case of longer storage term.

Impact

- Shows storage of long, non-combustible duration.
- Lessens wind and solar curtailment.
- The advocate of the dual carbon objectives of China.

Policy driver: LDES strategic investments and national energy storage roadmap.

8.7.7 Emerging trends in global deployment

Trend	*Description*
Shifting to long-duration storage	Increasing interest in flow batteries, CAES, hydrogen for >10-hour applications
Hybrid projects	Co-located solar + storage and wind + storage to optimize capacity and dispatch
Policy incentives	Governments offering tax credits, capacity payments, and tenders for storage
Declining costs	Lithium-ion prices dropped ~85% since 2010; new chemistries reducing cost further
Digital platforms	Use of AI and IoT to coordinate distributed storage and demand-side flexibility

8.7.8 Lessons from global leaders

i. **Certainty in policies is essential:** The fastest growth has been observed in markets that have clear targets and subsidies, and are regulated (e.g., California, Germany).

ii. **Technology diversity matters:** There is no one-size-fits-all system. Storage options must be local grid requirements, climate, and energy mix.
iii. **DERs and demand:** Storage is most complementary to solar, flexible loads, and digital control systems.
iv. **Finance and business models evolve:** Storage is becoming bankable as a service, leased, and stacked with many other value streams.
v. **Key enabling factors are the grid codes and standards:** Interoperability of storage and market access enhances investor confidence.

8.8 CHALLENGES AND BARRIERS TO STORAGE DEPLOYMENT

8.8.1 Introduction

Although energy storage is becoming widely recognized as an essential enabler of clean and flexible power systems, its ubiquitous deployment continues to experience numerous technical, economic, regulatory, and institutional impediments. These challenges should be understood to create policies, technologies, and business models that can be used to scale adoption. The section contains an elaborate description of the main barriers in the implementation of energy storage in the world and recommends their approaches to eliminate them.

8.8.2 Technical challenges

8.8.2.1 Technology maturity and performance

Although lithium-ion batteries are well developed, other technologies such as flow batteries, hydrogen storage, and CAES are in development or pilot. Key concerns include:

- Cycle life loss (e.g., lithium-ion cells have a capacity loss with time)
- Short shelf life of numerous chemistries (not more than 4 hours)
- RTE (losses of efficiency between 60% and 90%)
- Sensitivity to temperature, particularly of batteries in cold weather or hot climate.

8.8.2.2 Interoperability with grid infrastructure

- ESS should be interlocked with already modified grid protection, communication, and control systems. However, different manufacturers and grid operators do not have interoperability standards.
- High power requirement of electronic power and inverter control systems.
- Difficulty in co-location with generation or distribution property.

8.8.2.3 *Environmental and safety hazards*

- Fire in batteries, especially lithium-ion batteries, has been an issue. The threat of thermal runaway in improperly run battery banks.
- Toxic or flammable substances (e.g., electrolytes)
- Not having spent battery end-of-life recycling infrastructure.

8.8.3 Economic and financial barriers

8.8.3.1 *High upfront capital costs*

- ESS are expensive to set up, especially at grid scale: Cost of battery packs (but falling) is high.
- Power conversion systems, site development, and grid interconnection contribute to overall CAPEX.
- The payback periods are long, discouraging the private investors.

8.8.3.2 *Uncertain revenue streams*

In contrast to the conventional generators, storage systems do not generate energy but convert or store it, and this makes it harder to monetize:

- Value stacking (e.g., frequency regulation + energy arbitrage + peak shaving) is not necessarily available everywhere.
- Standardized contracts and predictable cash flows discourage investors.
- Models of revenue are different in different regions, making them risky.

8.8.3.3 *Poor financing facilities*

In most economies that are developing:

- Expensive debt and unavailability of concessional financing mechanisms.
- A lack of experience amongst banks and lenders in the analysis of storage projects.
- Minimal de-risking instruments (e.g., guarantees, insurance)

8.8.4 Regulatory and policy barriers

8.8.4.1 *Absence of regulatory classification*

In grid codes or energy acts in many countries, energy storage is not defined:

- Is it generation, load, or transmission? This influences licensing, connection charges, and access rights.
- There is regulatory confusion and delays as a result of ambiguity.

8.8.4.2 Market design limitations

Storage resources frequently have limitations in the engagement in energy and ancillary markets:

- None of the capacity markets, balancing reserves, and frequency control access.
- Operation is uneconomical in certain instances where there is a phenomenon of double charging (as a consumer and as a generator).
- No TOU pricing or dynamic tariffs diminish revenue ability.

8.8.4.3 Absence of long-term policy proxy

A lot of governments do not have a national roadmap of storage or uniform incentive systems:

- Pilot projects that do not have scale-up paths.
- Weak integration of renewable energy with storage.
- Shocks in policy (e.g., subsidy reductions) put investment off.

8.8.5 Social and institutional challenges

8.8.5.1 Community dynamics and accountability

- Low level of energy knowledge of storage and usefulness.
- Safety accidents (e.g., battery fires) can decrease local acceptance.
- Limitations to community involvement in the siting and ownership process.

8.8.5.2 Talent and labor gaps

- Professionally trained people required in battery system design, installation, working, and recycling.
- Lack of proper training programs, particularly in the developing economies.

8.8.5.3 Institutional coordination

- Disjointed duties among the utilities, regulators, and renewable energy agencies.
- Absence of centralized storage requirement and grid upgrades.
- Poor coordination between the industry and the research institutions.

8.8.6 Summary of key barriers

Category	*Major batteries*
Technical	Limited duration, degradation, safety, integration comp exity
Economic	High CAPEX, uncertain cash flows, limited access to finance
Regulatory	Undefined roles, limited market access, policy inconsistency
Social	Awareness gaps, safety concerns, workforce shortage
Institutional	Lack of planning, fragmented responsibilities, slow innovation transfer

8.8.7 Overcoming the barriers

In order to realize the potential of energy storage, it is necessary to address it through a multi-pronged approach:

- **Human technology development:** Invest in long-duration storage R&D, recycling, and thermal/hydrogen storage.
- **Market Reform:** Open up markets, eliminate the practice of double-charging, and make value-adding possible.
- **Policy incentives:** Provide capital subsidies, TOU pricing, and tax credits.
- **Finance innovation:** Concessional lending, blended finance, and project guarantees.
- **Capacity building:** This is to be achieved through the creation of training centers, certification programs, and public awareness campaigns.
- **Planning and integration:** Incorporate storage into integrated resource plans and grid expansion projects.

8.9 FUTURE OUTLOOK AND CONCLUSION

8.9.1 Future outlook

With the world in transition through the decarbonized, decentralized, and digitalized power system, the storage of energy is set to create transformational shifts in the way energy is generated, distributed, and consumed. Its capacity to absorb variability, balance supply-demand on-the-fly and to enhance system stability makes it the foundation of flexible grids.

8.9.1.1 Projected growth

The International Energy Agency (IEA) reckons that the global installed energy storage capacity (not counting pumped hydro) will increase 20-fold by 2030. The main factors in this geometric increase are:

- Increased national grid content of solar PV and wind.
- Extreme weather events that require backup are becoming increasingly common.
- Lower cost of batteries (should drop to less than 100/kWh).
- Transport and heating electrification.

8.9.1.2 Development of long storage

Although short-duration lithium-ion batteries are currently dominant, long-duration storage technologies, including the following, will be the way forward in the future.

- Flow batteries, with longer cycles and more scalable.
- Systems based on hydrogen, which allow seasonal storage as well as sector coupling are applied thermally.
- Thermal storage, particularly in industrial and district heating, mechanical storage such as gravity-based and compressed air.

These technologies are imperative to resolve the issue of multi-day or seasonal imbalances that accompany high renewable penetration.

8.9.1.3 AI and smart operation

The combination of AI, machine learning, and predictive analytics will enhance the operation of storage:

- Energy trading and frequency support optimization in real-time.
- Predictive health and maintenance.
- Smart scheduling of multi-service delivery.
- Enhanced storage asset value will be further increased by the use of digital twins as well as sophisticated forecast tools.

8.9.1.4 Policy and market evolution

Progressive energy markets will develop, leading to flexibility-first principles, including:

- Pricing that is time-varying in response to real-time grid conditions.
- Storage-and-solar-and-EVs energy-as-a-service models.
- The emergence of aggregators and VPPs into standard use.
- International standards that allow smooth trans-border operation.
- There will be a need to have global collaboration on the storage standards, safety, and recycling to ensure scalability and sustainability.

8.10 CONCLUSION

ESS have become an important technology in the world at large to decarbonize the electricity industry and, at the same time, maintain grid stability, reliability, and electricity flexibility. Storage offers a broad spectrum of technical and economic services at all grid levels, whether it is the optimization of renewable energy integration, or making customers involved, or contributing to resilient infrastructure.

Nonetheless, there are still major obstacles to overcome- technological maturity, economic viability, regulatory uncertainty, and lack of awareness. A systems-level solution that is inclusive of governments, utilities, regulators, investors, and consumers is needed to deal with these. An integrated effort of strategic investments, policies (inclusive), innovative ways of doing business, and capacity-building needs to converge to unlock the full potential of storage.

It will not be a simple grid that transmits and distributes electricity in the future, but rather it will store it, optimize it, and dynamically react to real-time situations. The energy storage of any type is essential to achieve the said vision.

REFERENCES

[1] International Energy Agency (IEA), *Electricity Market Report 2023*, Paris: IEA, 2023. [Online]. Available: www.iea.org

[2] U.S. Department of Energy, *Energy Storage Grand Challenge: Energy Storage Market Report*, Washington, DC, USA, 2021.

[3] A. Zakeri and S. Syri, "Electrical energy storage systems: A comparative life cycle cost analysis," *Renewable and Sustainable Energy Reviews*, vol. 42, pp. 569–596, Feb. 2015.

[4] M. A. Hannan, M. M. Hoque, A. Mohamed, and A. Ayob, "Review of energy storage systems for electric vehicle applications: Issues and challenges," *Renewable and Sustainable Energy Reviews*, vol. 69, pp. 771–789, Mar. 2017.

[5] International Renewable Energy Agency (IRENA), *Innovation Outlook: Smart Charging for Electric Vehicles*, Abu Dhabi: IRENA, 2019. [Online].

[6] Singh, P., Arora, K., Rathore, U. C., Joshi, G. P., & Cho, W. Comparative study of controllers in battery energy storage system integrated with doubly fed induction generator-based wind energy conversion system for power quality improvement. *Energy Reports,* vol. 11, pp. 4587–4600, 2024.

[7] G. Luo, S. Wang, Y. Li, and Z. Li, "Long-duration energy storage technologies: A review," *Applied Energy*, vol. 300, p. 117361, Oct. 2021.

[8] A. Nourai, *Installation of the First Distributed Energy Storage System (DESS) at American Electric Power (AEP): A Study for the DOE Energy Storage Systems Program*, Albuquerque, NM, USA: Sandia National Laboratories, 2007.

[9] M. G. Molina, "Energy storage systems for grid integration of renewable energy sources: A review of technologies, applications, and future trends," *IEEE Latin America Transactions*, vol. 14, no. 5, pp. 2119–2125, May 2016.

[10] European Commission, *Study on Energy Storage – Contribution to the Security of the Electricity Supply in Europe*, Brussels: Directorate-General for Energy, 2020.

[11] NITI Aayog and Rocky Mountain Institute, *India's Energy Storage Mission: A Market Access Pathway*, New Delhi, India, 2019.

[12] S. Yang, A. Kapoor, and T. Hossain, "A review of battery energy storage system integration with smart grid," *International Journal of Energy Research*, vol. 45, no. 14, pp. 21044–21064, Oct. 2021.

[13] U.S. Federal Energy Regulatory Commission (FERC), "Electric Storage Participation in Markets Operated by Regional Transmission Organizations and Independent System Operators," FERC Order No. 841, 2018.

[14] A. Akhil et al., *DOE/EPRI 2013 Electricity Storage Handbook in Collaboration with NRECA*, Albuquerque, NM, USA: Sandia National Laboratories, 2013.

[15] World Bank, *Energy Storage Partnership: Energy Storage Trends and Opportunities in Emerging Markets*, Washington, DC: World Bank Group, 2020.

Chapter 9

Clean and green energy

Advanced materials, technologies, and future perspectives

Prachi Saxena and Prashant

9.1 INTRODUCTION

The global imperative to combat climate change and transition away from fossil fuels has made clean and green energy a defining challenge of our time. This shift is not merely about adopting renewable sources like solar and wind power, but about fundamentally reimagining how we generate, store, and utilize energy. At the heart of this revolution lies a reliance on advanced materials and technologies that enhance performance, reduce costs, and improve the sustainability of energy systems. From cutting-edge perovskite solar cells to innovative catalysts for hydrogen production, the future of our energy landscape is inextricably linked to breakthroughs in materials science [1]. The journey toward a decarbonized energy system begins with generation, where advanced materials are revolutionizing the efficiency and accessibility of renewable technologies. In solar photovoltaics (PV), traditional crystalline silicon is being challenged by the emergence of perovskite materials. These compounds boast a higher power conversion efficiency, are made from abundant, low-cost materials, and are lightweight and flexible, opening up new applications in windows and wearable electronics. In wind energy, the efficiency of turbines is directly tied to the materials used in their construction. The use of advanced composites like glass and carbon fiber allows for the creation of longer, lighter blades. Additionally, research is focused on developing new alloys and magnet-free generator designs to reduce reliance on rare-earth materials like neodymium [2].

Beyond generation, the intermittency of renewable sources necessitates robust energy storage solutions. Advanced materials are central to the evolution of these technologies, particularly in the realm of batteries. While lithium-ion batteries dominate the market for electric vehicles and grid-scale storage, improvements are constantly needed in the materials for the cathode and anode to increase energy density and enhance safety. Looking to the future, solid-state batteries, which replace the flammable liquid electrolyte with a solid material, are a highly anticipated next-generation technology. For large-scale grid storage, flow batteries, which use abundant

DOI: 10.1201/9781003689430-9

elements like zinc and iron, are a cost-effective and scalable solution for long-duration energy storage [3]. The physical hardware of clean energy is only part of the solution; a smart, digital infrastructure is required to integrate these technologies into a cohesive and resilient system. Smart grids (SGs), for example, are a critical technology that uses digital communication to monitor and manage the electricity grid in real-time. This is enabled by the proliferation of Internet of Things (IoT) devices, such as smart meters and connected appliances, which provide the data streams necessary to optimize the grid's performance. Looking ahead, the future of clean energy presents both immense opportunities and significant challenges. One of the most promising future technologies is green hydrogen, produced by using renewable electricity to split water into hydrogen and oxygen through a process called electrolysis. Another future perspective involves carbon capture and storage (CCS) technologies, which capture carbon dioxide emissions from industrial processes or the atmosphere. However, a major challenge across all these technologies is the material supply chain, requiring a focus on sustainable sourcing and robust recycling programs [4]. In conclusion, the transition to a clean and green energy future is a testament to human ingenuity in materials science and technology. Advanced materials, coupled with a digitally interconnected grid, are enabling a new era of energy efficiency and resilience. While challenges such as material scarcity and cost remain, the rapid pace of innovation suggests that a future powered by clean energy is well within our reach.

9.2 MAJOR SOURCES OF CLEAN AND GREEN ENERGY

Clean and green energy refers to power derived from renewable resources that are naturally replenished and exert minimal adverse impact on the environment. Unlike fossil fuels, these energy systems do not release large quantities of greenhouse gases, making them essential for achieving climate neutrality and long-term sustainability [5]. The most widely recognized renewable energy sources (RESs) include solar, wind, bioenergy, hydropower with ocean energy, and hydrogen combined with fuel cell technologies. Solar power is regarded as one of the most promising renewable options due to the abundance of sunlight on Earth. It can be utilized in two major ways: directly converting sunlight into electricity or transforming it into heat [6].

9.2.1 PV systems

Solar cells, built with semiconducting materials, directly convert photons into electrical energy. *Monocrystalline panels* are cut from a single silicon crystal, offering superior efficiency and durability, though at a higher production cost. *Polycrystalline panels* are made from multiple silicon crystals, slightly less efficient but cheaper and widely available. *Perovskite-based PV* is an emerging alternative, valued for its low cost, lightweight structure, and ability to achieve high conversion rates. Solar thermal technologies [7]: These capture sunlight to generate heat, which can then be used for water heating, space heating, or electricity generation through steam turbines. Concentrated solar power (CSP) plants use mirrors or lenses to focus sunlight, reaching very high temperatures. Recent innovations: New designs such as *tandem solar cells* stack multiple layers to capture a wider range of the solar spectrum, pushing efficiency beyond conventional limits. *Quantum dot cells* introduce nanoscale materials with tunable properties, offering flexibility and lightweight applications. Integration with SGs and battery storage further enhances reliability [5].

9.2.2 Wind energy

Wind energy transforms the kinetic energy of moving air into mechanical power and subsequently into electricity. It has become one of the fastest-growing renewable sources worldwide.

Onshore wind farms are relatively inexpensive and easy to install, while offshore farms exploit stronger and steadier winds at sea, delivering greater output but requiring higher investment and complex infrastructure. Advances in Turbine Design: Modern turbines are taller with longer blades, enabling energy capture even at low wind speeds [6]. The use of lightweight composite materials, artificial intelligence for predictive maintenance, and aerodynamic improvements has significantly increased their efficiency. Hybrid applications: Wind farms are now being coupled with storage systems such as lithium-ion batteries or with complementary renewables like solar. Floating wind turbines are another innovation allowing deployment in deep waters where winds are more consistent [7].

9.2.3 Bioenergy

Bioenergy represents one of the most versatile and widely utilized forms of renewable energy. Unlike solar or wind energy, which depend heavily on weather conditions, bioenergy can provide a steady and controllable supply of power in the form of heat, electricity, and fuels. Derived from organic matter such as agricultural residues, forestry by-products, municipal waste, and dedicated energy crops, bioenergy contributes not only to energy generation but also to waste reduction, rural development, and climate change mitigation [5].

Globally, bioenergy accounts for a significant portion of renewable energy use, especially in rural and agricultural economies. Its flexibility in being converted into solid (biomass), liquid (biofuels), and gaseous (biogas) forms makes it essential for addressing diverse energy demands. However,

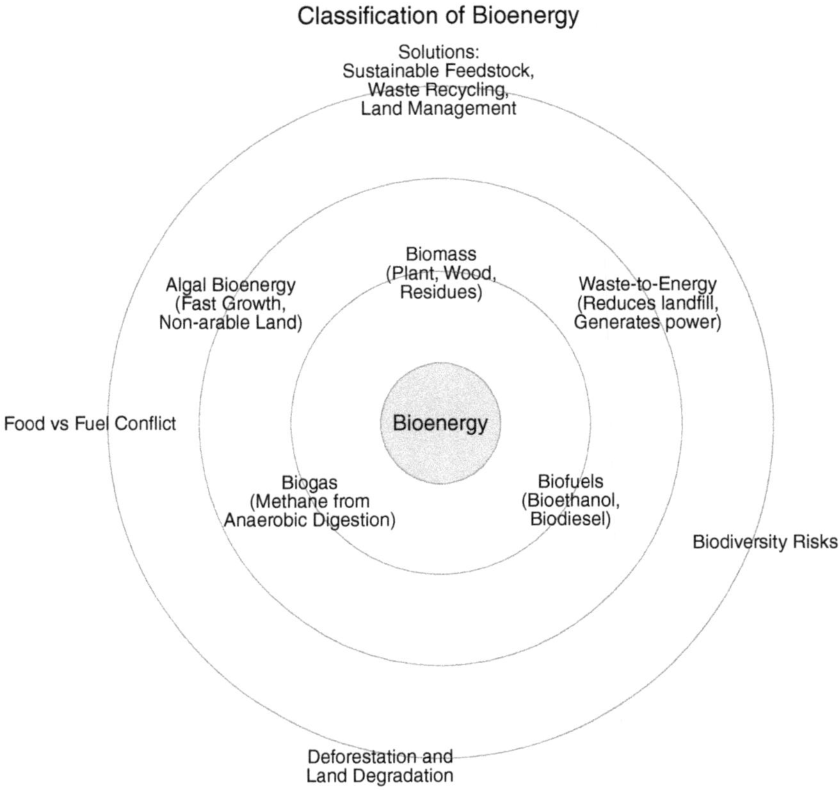

Figure 9.1 The classification in which various types of bioenergy have been presented [7–9].

questions of sustainability, land use, and environmental impacts must be addressed for its long-term success [6]. Figure 9.1 discusses the various forms of bioenergy used [7–9].

9.2.4 Hydropower and ocean-based energy

Hydropower and ocean-based energy systems form an important part of the global renewable energy portfolio. While hydropower has been utilized for centuries as a dominant renewable source of electricity, ocean energy technologies are relatively new and remain at different stages of research, demonstration, and commercialization. Together, they represent a critical pathway toward sustainable energy diversification. Hydropower is the process of converting the kinetic and potential energy of flowing water into electricity, primarily through the use of turbines and generators. It remains one of the largest contributors to renewable electricity globally, offering reliability and cost-effectiveness. Traditional large dams and reservoirs are capable of generating massive amounts of electricity and supporting grid stability [10]. They also provide multipurpose benefits such as flood control, irrigation, and water supply. However, such large projects are often associated with significant ecological and social impacts, including displacement of local populations, alteration of river ecosystems, and submergence of fertile land. Mini and micro-hydropower plants have emerged as sustainable alternatives, especially suited for rural and remote regions. Unlike large dams, these systems divert a portion of river flow without requiring massive reservoirs, thereby reducing environmental disturbances. Mini-hydropower (100 kW–1 MW) and micro-hydropower (<100 kW) are commonly used to provide decentralized electricity to off-grid villages. They are cost-effective, easy to maintain, and adaptable to local river conditions. Beyond electricity generation, they can be integrated with other renewables, such as solar or wind, to create hybrid rural energy solutions. Small-scale hydropower is, therefore, an essential enabler of rural electrification, supporting both community development and sustainability. Wave energy is derived from the motion of surface waves generated by winds across oceans. Devices such as point absorbers, oscillating water columns, and overtopping reservoirs are used to capture wave motion and convert it into electricity. It offers high energy density, making it attractive for coastal nations. However, the durability of devices in harsh marine environments and their high installation costs limit large-scale application at present [4].

9.2.5 Ocean energy systems

Oceans store immense quantities of energy, which can be harnessed in several forms. Although ocean energy technologies are still developing, they hold great promise for providing clean, predictable, and sustainable power. Tidal energy harnesses the natural rise and fall of ocean tides, driven by gravitational forces of the moon and sun. It is considered highly reliable because tidal cycles are predictable. Tidal barrages function like dams, capturing water during high tide and releasing it at low tide through turbines. Tidal stream turbines are underwater devices resembling wind turbines that rotate with tidal currents. Despite its predictability, tidal energy faces challenges related to high capital costs and potential impacts on coastal ecosystems and sediment flow [11]. Ocean thermal energy conversion (OTEC) exploits the temperature difference between warm surface water and cold deep-sea water. This temperature gradient is used to vaporize a working fluid, which drives a turbine to generate electricity before being condensed back by the cooler deep water. OTEC is particularly suitable in tropical and subtropical regions, where thermal gradients are most pronounced. Apart from power generation, it can also provide desalinated water and support aquaculture.

Nonetheless, low efficiency and high infrastructure costs continue to hinder its commercial-scale development [12].

9.2.6 Hydrogen and fuel cells

Hydrogen is increasingly seen as a clean carrier that can store, transport, and deliver energy without direct carbon emissions. Green hydrogen: When produced via electrolysis powered by renewables, hydrogen becomes a carbon-free fuel. Unlike grey hydrogen (from natural gas) or blue hydrogen (from gas with carbon capture), green hydrogen supports decarbonization goals. Fuel cells: Proton exchange membrane fuel cells (PEMFCs) are among the most developed, converting hydrogen into electricity with water as the only by-product is mentioned in Figure 9.2. They can be used in electric vehicles, portable devices, and backup power supplies. Hydrogen is critical for decarbonizing heavy industries such as steel, cement, and chemicals, where direct electrification is difficult. It also has potential for long-haul trucks, ships, and aircraft, as well as for storing excess renewable electricity at grid scale [13].

9.3 ADVANCED MATERIALS AND TECHNOLOGIES

9.3.1 Nanomaterials in energy conversion

The advancement of energy storage technologies for renewable sources is regarded as one of the most promising and auspicious ways to handle the present energy challenge and worldwide environmental issues. Investigating appropriate active materials constitutes one of the fundamental aspects of

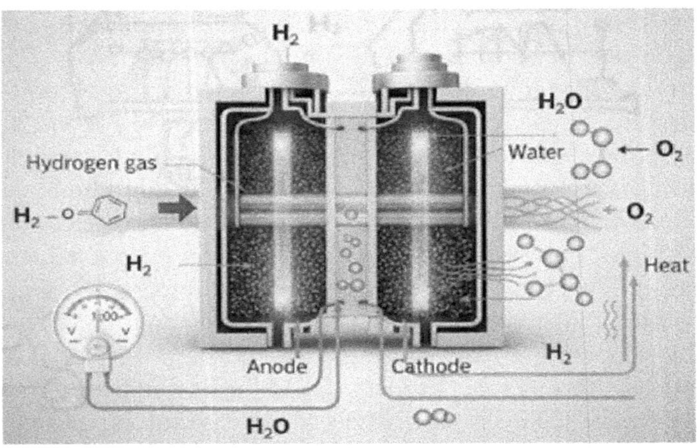

Figure 9.2 The presentation of working of hydrogen and fuel cells [13].

developing high-stable, high-efficient, environmentally friendly, and low-cost energy storage devices [14]. Next-generation energy technologies have seen nanomaterials emerge as pivotal components, particularly in the realm of batteries and energy materials. Various energy applications have garnered significant attention from inorganic nanomaterials due to their unique thermal, mechanical, optical, and electrical properties. However, addressing the challenges posed by scaling relationships within energy devices and inorganic nanomaterials is imperative to fully harness their potential [13, 15]. Important roles are played by inorganic multifunctional nanomaterials in energy storage, energy generation, energy saving, energy conversion, and energy transmission applications due to their distinctive properties such as chemical stability, higher surface area, and outstanding thermal and electrical conductivity. A path is being developed for future energy-related applications using inorganic multifunctional nanomaterials characterized by lower toxicity, lower cost, more functions, and higher performance. Despite their terrific potential, significant efforts are still needed by the scientific research community to address their execution and achievement in extensive implementations. The mechanism of working of inorganic multifunctional nanomaterials remains fuzzy [15, 16]. Researchers suggested nanomaterials with a high specific surface area to direct future investigation towards reducing the difference between the discovered and predicted capacitance without restricting the loading mass. Batteries and supercapacitors are two types of storage devices that store electrochemical energy, which are crucial for enabling intermittent yet sustainable energy alternatives like solar and wind. Supercapacitors can achieve high power because they exclusively store charge at their surfaces, where diffusion processes do not hinder charge storage. Similarly, supercapacitors are significantly more reversible due to their longer life cycle, as charging/discharging does not require a bulk-phase shift/transformation [16, 17]. Batteries generally store charge in the bulk phase of electrode materials through redox reactions, resulting in advanced energy density but lower power output compared to supercapacitors. Recently, the desire for high-efficiency, resourceful batteries has become ubiquitous and tangible, with the numerous advantages and demands of these batteries almost becoming common knowledge [17]. A class of materials or material-based systems known as nanomaterials is represented by those designed to strategically combine multiple properties to fulfil various functions. Efficiency at the system level that surpasses that of individual components is aimed to be achieved by these materials. Generally, the same length scale as the multifunctional characteristics of materials can be applied to their functionality, either concurrently or subsequently. Unique electrical, magnetic, optical, and other properties are encompassed by these remarkable qualities, which can significantly impact fields such as spintronics, electronics, medical applications, and technological advancements [16, 17]. A vast materials class with substantial promise for energy transmission applications is represented by multifunctional inorganic nanomaterials, with a primary focus on superconducting materials and smart electricity networks. The goal of realizing a room-temperature superconductor has been pursued by scientists ever since the discovery of superconductivity. A variety of unusual physical and electrical phenomena for superconductors, including 2D superconductivity and a decrease in superconducting transition temperature as sample thickness is reduced to the nanoscale, have been revealed by recent experimental developments on inorganic nanomaterials [18].

9.3.2 Energy storage systems

The development of advanced energy storage systems has been necessitated by the rapid growth of RESs like solar and wind to address the intermittent nature of these sources. Energy storage systems are significantly enhanced by nanotechnology through various mechanisms like increased surface area, improved charge transport, and electrode stability [18]. Compared to bulk materials, nanostructures like nanowires, nanotubes, and nanoparticles possess a much higher surface area, which significantly improves energy storage performance. More active sites for energy storage reactions are provided by this fact, resulting in higher energy densities as well as faster rates of charging and discharging [19]. Charge transport within energy storage devices is also improved by the unique properties of nanomaterials, boosting the efficiency and performance of batteries and supercapacitors. However, a significant challenge in current research remains in effectively and efficiently utilizing stored energy and power. Well-known energy storage devices include supercapacitors (SCs), batteries, and fuel cells. The gap between capacitors and batteries is bridged by SCs, offering a balance between the two technologies. Commonly utilized as electrodes in EDLCs are carbonaceous materials such as graphene, reduced graphene oxide, carbon nanotubes (CNTs), and activated carbon [19, 20]. Energy is stored through faradaic redox reactions by pseudocapacitive materials, which represent a different class of nanomaterials. Employed as electrodes in pseudo capacitors are various substances like conducting polymers, metal oxides, hydroxides, etc. Different mechanisms for energy storage are offered by both EDLCs and pseudo capacitors through the utilization of these distinct electrode materials, providing a range of options for SC technology [20]. 2D and 3D nanomaterials are well recognized to have great potential as high-performance electrode materials for energy storage systems. Improvements in bandgap, electrical conductivity, structural flexibility, and charge carrier mobility are required in the electrochemical charge storage mechanism of SCs to resolve problems. Particularly, the exposure of active sites is maximized, volume fluctuations during reactions are buffered, and charge-carrying ion diffusion paths are shortened by the extensible planar structure of 2D nanomaterials with ultrathin thickness [21, 22]. These unique features make these nanomaterials highly desirable for energy storage applications. However, several challenges are linked to the scalable production of high-quality 2D nanomaterials, particularly for environmentally sensitive ones. The performance of 2D nanomaterials during electrode manufacturing and electrochemical reactions is hindered by issues such as layer restacking, aggregation, and surface chemical inertness. The full potential of the materials is prevented from being reached by these limitations [22]. To overcome these challenges and to enhance the charge storage capacity and efficiency, various functionalization strategies have been proposed, including heteroatom doping, defect engineering, heterostructure formation, and pore-construction. The electrochemical performance of the nanomaterials is aimed to be enhanced by these strategies through the improvement of their chemical affinity to target ions and addressing other intrinsic problems. The development of novel nanomaterials and electrodes with exceptional electrochemical performance has been led by the implementation of these functionalization strategies, thus advancing the field of energy storage systems [23].

9.3.3 SGs and digital energy systems

Various RESs such as solar PV systems, wind systems, and small hydro systems are made use of by the microgrid (MG) and nano grid (NG) systems. Many loosely interconnected synchronous alternate current (AC) grids make up a traditional power grid [23]. The grid carries out three main functions: generating, transmitting, and distributing electrical energy, where electric power flows only in one direction, from a service provider to the consumers [24]. A significant wastage of energy is endured by the power grids due to several factors, such as inefficient appliances used by consumers and the lack of smart technology, inefficient routing and dispensation of electrical energy, unreliable communication and monitoring, and, most importantly, the lack of a mechanism to store the generated electrical energy [23, 24] has been described in Figure 9.3. Furthermore, some other challenges are faced by the power grids as well, including growing energy demand, emerging RESs, reliability, security, and aging infrastructure problems, to name a few. To solve these challenges, researchers have identified the SG paradigm as a promising solution that uses a variety of information and communication technologies [24]. These technologies can enhance the effectiveness, sustainability, stability, efficiency, reliability, security, and scalability of the traditional power grid. Traditional power grids differ from SG in many aspects. For instance, a bi-directional communication flow is offered by SG between service providers and consumers, while uni-directional communication is only offered by a traditional power grid from the service provider to the consumer [7].

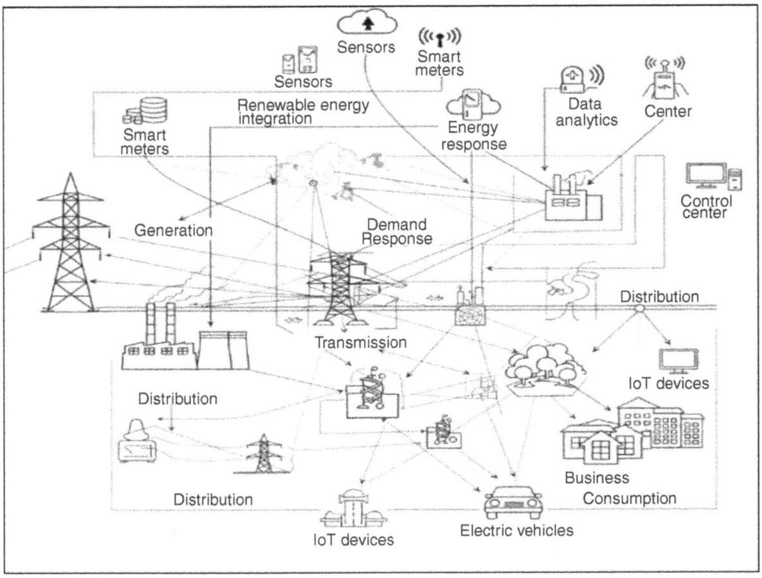

Figure 9.3 Smart grids and digital energy systems [23–25].

The problem of electrical energy wastage is solved by SG through the generation of electrical energy that closely matches the demand. Important decisions, according to the demand for energy, such as real-time pricing, self-healing, power consumption scheduling, and optimized electrical energy usage, are helped to be made by SG [24, 25]. One of the main concerns for SG is the connectivity, automation, and tracking of many devices, which necessitates distributed monitoring, analysis, and control through high-speed, ubiquitous, and two-way digital communications [27]. Distributed automation of SG for such devices or "things" is required. The real world is already realizing this through the IoT technology. Researchers consider SG to be one of the largest applications of IoT. By establishing IoT technology as a global standard for communications and the foundation for SG, we will open new doors for maximizing the prospects for future innovations [23–28].

9.4 ENVIRONMENTAL AND SOCIOECONOMIC IMPACTS

As the world's population reaches 8 billion, rising energy consumption and production activities result in most of the energy consumption being sourced from non-renewable sources [29]. The adverse effects of the perpetual use of non-renewable energy are causing global warming to impact the environment. Globally, researchers consider the combustion of fossil fuels to be the highest emitter of greenhouse gases [30]. The environmental consequences of emissions have prompted the world to focus on a universal goal, thus driving the fight to replace fossil fuels with sustainable energy. Irrespective of the role fossil fuels play in the development of the economy in many countries, especially developing countries, environmentalists, government bodies, non-state actors, and all concerned individuals have put measures in place to replace it due to the threat it poses to the environment [31, 32].

Over the last decade, leading renewable energy technologies such as wind power and solar PV have experienced rapid or substantial cost declines [33]. In 2021, renewable energy sectors worldwide employed at least 12.7 million people, both directly and in wider supply chains, with most of these jobs concentrated in China, Brazil, the EU, the United States, and India. Considering the economic implications of the COVID-19 pandemic and recent gas and electricity price spikes [34]. Various stakeholders have made calls for investment in green jobs, skills, and infrastructure to help consolidate economic recovery from COVID-19 and the greenhouse effect in a way that is compatible with achieving net-zero emissions and a societally just transition [35, 36]. A key definitional issue is the distinction between "gross" and "net" jobs. Only the positive impacts on employment associated with a particular investment are included in gross effects. For any specific energy technology, gross jobs involve those created from project spending on equipment manufacturing, installation, operation, and maintenance [36]. Lower-skilled, manual occupations are also in demand, comprising significant shares of solar PV installation and offshore wind construction activities, for example. Nevertheless, several studies suggest that job losses will most likely affect lower-skilled workers because of low-carbon energy transitions. Stronger labor market policies and regulations are needed to ensure displaced employees can access alternative, decent work in low-carbon sectors [37].

The deployment of large capacities of renewable energy technologies is required by the transition toward a global carbon-neutral energy system. However, a wide variety of raw material demands incomparable to those associated with fossil-fuel-based energy systems are introduced by these technologies [38]. These raw material demands have already noticeably increased and will be raised even higher over the next decades. Many of the required materials are at risk of facing supply disruption because high costs and geopolitical complications are associated with their extraction. Therefore, an increasing amount of attention has been drawn to the so-called energy-material nexus [38, 39].

9.5 CHALLENGES AND LIMITATIONS

Environmental concerns arise nowadays due to the emission of carbon dioxide, sulfur dioxide, and nitrogen oxides from fossil fuel-based generating stations. Global warming and acid rain are caused by this environmental pollution. In contrast, traditional synchronous generator-based power generation is compared with renewable energy (RE) generation systems, which are cleaner and cheaper. Consequently, the renewable energy generation is increased by the governments and several agencies to replace fossil fuel-based power generation. Nowadays, high-level integration of converter-based RESs yields several economic and technical benefits, such as low-cost energy, reduced carbon emissions, and lower operational and maintenance costs. However, high-level renewable integration raises many technical issues, including very low inertia that causes frequency instability, high fault currents resulting from short circuits, intensive uncertainties due to varying wind speeds and irradiance, and degraded power quality. Researchers, system planners, and operators find it challenging to maintain flexible, reliable, and stable operations of such systems. However, researchers use and continuously develop several cutting-edge technologies and techniques to address new challenges resulting from renewable energy integration. For instance, they adopt virtual inertia control, fault current limiters, advanced filters, energy storage devices, and optimization techniques to handle those issues. Nevertheless, possibilities still exist to develop better strategies for addressing the various challenges of high-level renewable energy integration. Advanced control methodologies should consider high power rating of converters and improved power sharing among the converters to facilitate high-level RES integration through redesign, improvement, or implementation [39]. Energy storage devices play an important role in supporting frequency and voltage profiles. However, few researchers examine how the lifespan of energy storage systems affects their application for frequency and voltage support, power quality enhancement, and uncertainty management.

REFERENCES

1. Jain, H. (2024). From pollution to progress: Groundbreaking advances in clean technology. *Heliyon*, 10(5), e07755. https://doi.org/10.1016/j.heliyon.2023.e07755
2. Sharma, V. K. (2024). A comprehensive review of green energy technologies. *MDPI Energies*, 13(1), 69. https://doi.org/10.3390/en13010069
3. Sundaram, G. A., & Muniyandi, G. R. (2025). Advancements in renewable energy: innovations and sustainable solutions for a greener future. *Next Research*, 2(3), Art. no. 100482. https://doi.org/10.1016/j.nexres.2025.100482.
4. Tiwari, A. (2024). How advanced materials drive our net-zero future. *Advanced Materials Letters*, 15(4), 24472. https://doi.org/10.5185/amlett.2024.24472
5. Finecomess, S. A. (2024). Future green energy: A global analysis. *MDPI Energies*, 17(12), 3039. https://doi.org/10.3390/en17123039

6. Engel-Cox, J. A., Merrill, W. G., Mapes, M. K., McKenney, B. C., Bouza, A. M., DeMeo, E., Hubbard, M., Miller, E. L., Tusing, R., & Walker, B. J. (2022). Clean energy technology pathways from research to commercialization: Policy and practice case studies. *Frontiers in Energy Research*, 10, Art. no. 1011990. https://doi.org/10.3389/fenrg.2022.1011990.
7. Okeke, G. N. (2025). An evaluation of renewable energy technologies and their applications in electricity generation. *Renewable and Sustainable Energy Reviews*, 15(7), 1234–1245. https://doi.org/10.1016/j.rser.2025.03.015
8. Cheema, H. A. (2024). Comprehensive review of the global trends and future directions in photovoltaic recycling. *Science of the Total Environment*, 836, 155531. https://doi.org/10.1016/j.scitotenv.2022.155531
9. Jayabal, R. (2024). Innovations in green energy for a sustainable future. *Nature Sustainability*, 7(1), 1–15. https://doi.org/10.1038/s41599-023-01901-3
10. Sharma, V. K., Monteleone, G., Braccio, G., Anyanwu, C. N., & Aneke, N. N. (2025). A comprehensive review of green energy technologies: Towards sustainable clean energy transition and global net-zero carbon emissions. *Processes*, 13(1), 69. https://doi.org/10.3390/pr13010069
11. Wheatley, M. C. (2024). Advancements in renewable energy technologies. *Premier Journal of Science*, 24(256), 1–15. https://premierscience.com/wp-content/uploads/2024/11/pjs-24-256.pdf
12. Sharma, P., Verma, P., Reddy, K. R., & Kim, K. H. (2025). Seeding the future of nanomaterials: A comprehensive review of nanosheet-mediated growth for energy harvesting, energy conversion, and photodetection applications. *RSC Advances,* 15(23), 16123–16143. https://doi.org/10.1039/D5RA01655J
13. Kumar, A., Singh, R., & Zhao, H. (2024). Nanomaterial-based energy conversion and energy storage devices: A comprehensive review. *New Journal of Chemistry,* 48(3), 1246–1281. https://doi.org/10.1039/D3NJ04846B
14. Zhang, Y., Wang, J., & Chen, X. (2024). Advances in nanomaterials for energy conversion and environmental catalysis. *Nanomaterials,* 14(23), 1902. https://doi.org/10.3390/nano14231902
15. Ali, S., Patel, R., & Gupta, D. (2024). Photocatalytic nanomaterials and their implications towards biomass conversion for renewable chemical and fuel production. *Nanoscale Advances,* 6(15), 4055–4072. https://doi.org/10.1039/D4NA00447G
16. Li, J., Zhou, Y., & Huang, P. (2024). Transition metal nanomaterial-based electrocatalysts for water and CO_2 electrolysis: Preparation, catalytic activity, and prospects. *Frontiers in Energy Research,* 12, 1433103. https://doi.org/10.3389/fenrg.2024.1433103
17. Lamba, P., Singh, P., Singh, P., Singh, P., Bharti, Kumar, A., Gupta, M., & Kumar, Y. (2022). Recent advancements in supercapacitors based on different electrode materials: Classifications, synthesis methods and comparative performance. *Journal of Energy Storage*, 48, 103871. https://doi.org/10.1016/j.est.2022.103871
18. Alonzo, S. M. M., Bentley, J., Desai, S., & Bastakoti, B. P. (2023). Hydrothermal synthesis of hierarchical microstructure tungsten oxide/

carbon nanocomposite for supercapacitor application. *Scientific Reports*, 13, 21732. https://doi.org/10.1038/s41598-023-48972-3

19. Conway, B. E., Birss, V., & Wojtowicz, J. (1997). The role and utilization of pseudocapacitance for energy storage by supercapacitors. *Journal of Power Sources*, 66(1), 1–14. https://doi.org/10.1016/S0378-7753(97)02025-7
20. Yang, X., Zhang, L., Zhang, F., Zhang, T., Huang, Y., & Chen, Y. (2014). A high-performance all-solid-state supercapacitor with graphene-doped carbon material electrodes and a graphene oxide-doped ion gel electrolyte. *Carbon,* 72, 381–386. https://doi.org/10.1016/j.carbon.2014.01.061
21. Liu, T., Ding, J., Su, Z., & Wei, G. (2017). Porous two-dimensional materials for energy applications: Innovations and challenges. *Materials Today Energy,* 6, 79–95. https://doi.org/10.1016/j.mtener.2017.02.002
22. Tondan, H., & Singh, A. K. (2024). Advances in energy harvesting and storage materials: Unlocking the potential of solid-state nanomaterials for renewable energy technologies. In *Futuristic trends in physical sciences*, Volume 3, Book 4 (pp. 21–32). Iterative International Publishers, Selfypage Developers Pvt Ltd. ISBN 978-93-5747-671-3
23. Collier, S. E. (2017). The emerging Enernet: Convergence of the Smart Grid with the Internet of Things. *IEEE Industry Applications Magazine*, 2, 12–16. https://doi.org/10.1109/MIAS.2017.7901123
24. Deng, R., Yang, Z., Chow, M.-Y., & Chen, J. (2015). A survey on demand response in smart grids: Mathematical models and approaches. *IEEE Transactions on Industrial Informatics*, 11(3), 570–582. https://doi.org/10.1109/TII.2015.2428819
25. Ma, R., Chen, H.-H., Huang, Y.-R., & Meng, W. (2013). Smart grid communication: Its challenges and opportunities. *IEEE Transactions on Smart Grid*, 4(1), 36–46. https://doi.org/10.1109/TSG.2012.2218497
26. Singh, P., Arora, K., Rathore, U. C., Joshi, G. P., & Cho, W. (2024). Comparative study of controllers in battery energy storage system integrated with doubly fed induction generator-based wind energy conversion system for power quality improvement. *Energy Reports,* 11, 4587–4600. https://doi.org/10.1016/j.egyr.2024.03.219
27. Bush, S. (2014). Network theory and smart grid distribution automation. *IEEE Journal on Selected Areas in Communications*, 32(7), 1451–1459. https://doi.org/10.1109/JSAC.2014.2323431
28. Wang, C., Li, X., Liu, Y., & Wang, H. (2014). The research on development direction and points in IoT in China power grid. In International Conference on Information Science, Electronics and Electrical Engineering (ISEEE) (Vol. 1, pp. 245–248). IEEE. https://doi.org/10.1109/ISEEE.2014.6994445
29. Sarkodie, S. A., Adams, S., Owusu, P. A., Leirvik, T., & Ozturk, I. (2020). Mitigating degradation and emissions in China: The role of environmental sustainability, human capital and renewable energy. *Science of the Total Environment,* 719, 137530. https://doi.org/10.1016/j.scitotenv.2020.137530
30. Xu, L., Wang, X., & Guo, W. (2022). Does renewable energy adaptation, globalization, and financial development matter for environmental quality and economic progress? Evidence from panel of big five (B5)

economies. *Renewable Energy*, 192, 631–640. https://doi.org/10.1016/j.renene.2022.04.123
31. Anusha, G., Arora, K., Sharma, H., Thota, S. P., Joshi, G. P., & Cho, W. (2024). Control strategies of 15-level modified cascaded H-bridge MLI with solar PV and energy storage system. *Energy Reports,* 12, 2–26. https://doi.org/10.1016/j.egyr.2023.11.172
32. Stern, N., & Valero, A. (2021). Innovation, growth and the transition to net-zero emissions. *Research Policy*, 50(9), 104293. https://doi.org/10.1016/j.respol.2021.104293
33. Acar, S. et al (2023). Transforming Türkiye's power system: An assessment of economic, social, and external impacts of an energy transition by 2030. *Clean Energy Systems*, 4, 100064. https://doi.org/10.1016/J.CLES.2023.100064
34. Aceleanu, M. I., Serban, A. C., & Burghelea, C. (2015). "Greening" the youth employment—A chance for sustainable development. *Sustainability*, 7(3), 2623–2643. https://doi.org/10.3390/su7032623
35. Allan, G. J., & Ross, A. G. (2019). The characteristics of energy employment in a system-wide context. *Energy Economics* 81, 238–258. https://doi.org/10.1016/J.ENECO.2019.03.017
36. Arvanitopoulos, T., & Agnolucci, P. (2020). The long-term effect of renewable electricity on employment in the United Kingdom. *Renewable and Sustainable Energy Reviews*, 134, 110322. https://doi.org/10.1016/j.rser.2020.110322
37. Kowalski, P. & Legendre, C. (2023). Raw materials critical for the green transition: Production, international trade and export restrictions. *OECD Trade Policy Papers*, no. 269, OECD Publishing, Paris, France, pp. 1–93. https://doi.org/10.1787/c6bb598b-en.
38. Buchholz, P., & Brandenburg, T. (2018). Demand, supply, and price trends for mineral raw materials relevant to the renewable energy transition wind energy, solar photovoltaic energy, and energy storage. *Chemie Ingenieur Technik*, 90, 141–153. https://doi.org/10.1002/cite.201700098.
39. Alam, M. S., Al-Ismail, F. S., Salem, A., & Abido, M. A. (2020). High-level penetration of renewable energy sources into grid utility: Challenges and solutions. *IEEE Access,* 8, 190277–190299.

Chapter 10

Integration of grid with distributed energy resources

Technologies, potential contributions, and future prospects

Shahid Shabir, Krishan Arora, and Javed Dhillon

10.1 INTRODUCTION

The increasing demand for clean, efficient, and reliable electricity has led to the integration of distributed energy resources (DERs) into the power grid [1]. Traditional power grids faced a lot of challenges, which led to the end of traditional power grids, and DERs came into place. The International Energy Agency (IEA) estimates that until 2030, the world's demand for electrical energy would grow at a steady 1.4% yearly rate [2, 3]. Traditional power systems, which depend on centralized grid topologies with a few large-scale energy producers and countless end consumers, are under a lot of strain due to this rising consumption. However, a paradigm shift toward more robust and adaptable power distribution systems is required because such centralized arrangements are vulnerable to failures, operational inefficiencies, and financial losses. It is now essential to convert passive distribution networks into actively managed systems to overcome these difficulties. Adoption of cutting-edge commercial frameworks and technology solutions that facilitate decentralized power generation is necessary for this shift [4]. Distributed generation (DG), which refers to small-scale electricity generation units that can connect at different places within the power grid, is one of the most promising solutions. The Institute of Electrical and Electronics Engineers (IEEE) claims that DG includes energy production from localized sources, enabling increased system efficiency and dependability [5].

DERs, including fuel cells (FCs), wind turbines, and solar photovoltaic (PV) systems, are anticipated to be essential to the world's energy production in the context of future smart grids [6]. Enhanced grid resilience, decreased transmission losses, and greater sustainability are just a few advantages of these decentralized energy sources. Nevertheless, there are a number of difficulties with their extensive integration, such as grid stability, voltage regulation, and the requirement for sophisticated management and protection systems [7]. With an emphasis on solar PV, wind power, and FC technologies, this chapter investigates the technical and commercial integration of

 DOI: 10.1201/9781003689430-10

DERs within contemporary distribution networks [8]. Three main points are highlighted in the study: the economic impact and commercial viability of DER adoption. The benefits and technical difficulties of integrating DER [9]. Future research and development (R&D) must improve grid interoperability and DER deployment. By tackling these issues, the study hopes to shed light on how DER is changing in contemporary power networks and what innovations are required to guarantee a steady and sustainable supply of electricity [10].

10.2 DISTRIBUTED GENERATION

10.2.1 Literature review

Although the idea of DG is not new, it has drawn more attention recently as a potential way to improve the efficiency and dependability of the power system [11]. The word "distributed" refers to a range of generation scenarios that are influenced by a number of important factors, such as government initiatives to promote sustainable energy production and lower CO_2 emissions [12]. Developments in digital communication networks, energy storage, and renewable energy sources. Changes to regulations that promote innovation and competition in the energy industry [13]. Power generation is moving from massive, centralized power plants to smaller, dispersed units situated closer to users as a result of the move toward decentralized energy systems [14]. Numerous benefits come with this shift, such as enhanced energy security, decreased environmental impact, and higher system efficiency. There is not a single, widely recognized definition of DG in the literature, despite its increasing importance [15]. Definitions frequently differ according to ownership, technology, location, scale, impact on the environment, and purpose [16]. The definition of DG by the IEEE is "the generation of electricity by facilities sufficiently smaller than central plants, usually 10 MW or less, allowing interconnection at nearly any point in the power system" [17]. This description emphasizes how DG may be deployed in a variety of ways and how it can improve grid resilience. The function of DG in contemporary power systems has been the subject of numerous studies, with an emphasis on the technical, financial, and environmental ramifications [18]. According to research, DG integration into distribution networks can improve grid stability, lower transmission losses, and pave the way for a more sustainable energy future [19]. To guarantee smooth DG integration, however, issues including protection coordination, power quality control, and voltage regulation must be resolved [20]. To optimize deployment techniques, evaluate its impact on power system dynamics, and create regulatory frameworks that facilitate its widespread adoption, more study is required as DG continues to advance [21].

10.2.2 Benefits of DER

1. Advantages for the environment: DER encourages the efficient and widespread use of fossil energy resources (FERs) [22]. By using combined heat and power (CHP) systems, environmental thermal pollution is reduced. Energy-generating-related emissions can be reduced with the use of DG.
2. Benefits compared to centralized power facilities: DER makes it possible for several sources to share power generation. It gives you the freedom to work on your own. By improving local control, power systems can better adjust to local needs.
3. Effects on the energy industry: Ancillary services are something that DER contributes to. It increases competition in the energy industry, which promotes cost-cutting and efficiency.
4. Financial benefits: By reducing or doing away with the need for extra transmission and distribution (T&D) infrastructure, DER lowers planning and expansion costs. Modular DGs are simple to install and can be placed in a variety of locations, especially in places where renewable energy sources need certain geographic circumstances. By using waste heat for heating, cooling, or additional power generation, CHP-based DGs improve overall efficiency. By providing power to the grid and lowering reliance on high-demand production, the availability of DGs can contribute to a decrease in wholesale electricity prices. In addition to saving fuel, they help to extend the life and effectiveness of system parts, such as transformers.
5. Improvement of security in the power system: In distribution networks, DER improves electricity quality and voltage profiles. It lowers energy losses in the distribution and transmission networks. In demand response tactics like "peak load shaving" and load control initiatives, DGs are essential. For end users, the stability and dependability of the power supply are strengthened by the inclusion of DER. DGs can serve as on-site backup power sources during emergencies or outages. They ensure a smoother flow of electricity by easing transmission limitations. To maintain system stability, DER lessens the requirement for extra spinning reserves. Increased DG adoption helps reduce power grid congestion, which eventually improves transmission networks' efficiency.

10.2.3 Barriers of DER

1. Centralized power plant challenges include: Since DER information is frequently dispersed, integrating it into current power systems is challenging. Operational challenges arise from DER implementation's

lack of uniformity and standards. To transition to DER-based power generation, a lot of administrative and educational work is needed.

2. Obstacles in the energy sector: Appropriate incentive schemes are required to promote DER technology integration and uptake.
3. Issues with power system security: Bidirectional power flows in distribution networks require the development or modification of current control systems. Because gas supply and demand fluctuate, the growing usage of natural gas-powered CHP systems in distribution grids raises questions about supply security. The traditional power flow is changed by connecting DG technology, which affects fault current levels and local voltage stability. One unforeseen consequence of integrating DG technology into the grid through power electronic converters could be a decline in power quality. Resolving grid imbalances will be more important to preserving system stability as heat-based CHP systems and intermittent renewable energy sources gain market dominance.

The International Council on Large Electricity Systems (CIGRE) describes a DG unit as a power generation source that is not part of central planning or dispatch, typically linked to the distribution network, and generally has a capacity below 50–100 MW. Similarly, the IEA defines DG as a generation facility that supplies power directly to an on-site customer or supports the distribution network, connecting at distribution-level voltages. According to the U.S. Department of Energy (DOE), DG refers to compact electric generation or storage systems positioned close to the point of consumption. The DOE further classifies distributed power systems as those typically under 1 MW, including energy storage solutions, located near consumer loads or at distribution and sub-transmission substations, collectively termed DERs.

The European Directive 2003/54/EC defines DG as "generation plants connected to the distribution system." Based on various definitions, DERs can be summarized as electricity generation sources linked to the grid at distribution-level voltages, either supplying power directly to an on-site consumer or supporting the distribution network. Furthermore, as an alternative to conventional energy systems, distributed energy systems offer a more efficient, reliable, and environmentally sustainable approach. To fully understand the advantages of DG, it is essential to evaluate its impact on the environment, power system performance, efficiency, economics, and energy security. A concise overview of its benefits and limitations is provided in "Benefits of DER" and "Barriers of DER" sections, respectively.

10.3 DG CLASSIFICATION

The intermittent nature of some DG systems can be managed with the use of energy storage technologies like flywheels and batteries. DG units can

Table 10.1 DG classification criteria

Classification criteria	*Description*
Capability for emission-free operation	Whether the generation unit operates without emissions (e.g., renewables like solar, wind).
Type of energy source or prime mover	The source of energy used (e.g., fossil fuels, renewables) or the type of engine/turbine driving the generation.
Suitability for CHP production	Whether the system can produce both electricity and heat efficiently.
Available unit size and modularity feasibility	The range of unit capacities and the possibility of scaling by adding more units.
Intermittent power output	Whether the power generation is continuous or variable (e.g., solar and wind are intermittent).
Type of dispatch (local and/or central)	Whether the generation is managed locally or integrated into the central grid dispatch system.

Table 10.2 Most common classification of DG capacities

DG capacity range	*Category*
1 W–5 KW	Micro
5 KW–5 MW	Small
5 MW–50 MW	Medium
50 MW–300 MW	Large

be categorized according to a number of criteria, such as their electrical uses, power supply duration, electricity generation type, capacity ratings, and whether or not they use renewable or non-renewable energy sources. Figure 10.1 shows the most commonly used classification. From a single unit to several units joined in a modular arrangement, these capacity levels can vary greatly (Tables 10.1 and 10.2).

A prospective from micro-CHP systems that use Stirling engines, FCs, and microturbines to renewable energy sources like solar PV systems, wind energy conversion systems (WECS), and small-scale hydroelectric power generation, DER technologies come in a variety of forms. DG can also be grouped according to the fuel type, which can be classified as either fossil-based or non-fossil-based energy sources. As was previously said, DG systems can be used in a variety of contexts, each with unique benefits and difficulties that need careful consideration. The following DER technologies are the main topic of this paper: WECS, solar PV, and FC

Figure 10.1 Wind turbine.

10.4 DISTRIBUTED ENERGY RESOURCES

10.4.1 Systems for converting wind energy

Wind energy is converted into electrical power using WECS.

The wind turbine, which uses rotor blades to capture wind energy, is the essential part of a WECS. A gearbox then transfers this energy to the induction generator. The generator shaft is driven by the wind turbine, which generates power. There are two possible configurations for wind turbines: horizontal and vertical. In WECS, commercial wind turbines usually have capacities between 300 kW and 5 MW. The main applications for these systems are in industrial operations and remote locations. WECS are considered variable power sources because of their weather-dependent operation and the need for control in some applications.

In addition to DERs' overall benefits, WECS are a green technology. Unlike nuclear power, they do not produce CO_2 while in operation, improve global safety by avoiding radioactive or hazardous waste, and, unlike conventional oil-based fuels, their costs are continuously decreasing.

10.4.2 Wind energy's obstacles in DG

- Wind energy adoption is hampered by a number of issues in different market segments: The limited availability and range of turbine sizes are a significant drawback. In addition to inadequate turbine efficiency and design, the lack of commercially available turbines weighing 50 kW or more results in missed market possibilities.
- Standards and needs for grid interconnection: Connecting wind turbines to the grid is a complicated procedure, with different utility providers and regions having different technical needs and laws.
- Grid integration's technical and financial challenges: The integration of small wind systems into the electrical distribution network presents both financial and technical challenges, requiring additional developments in system optimization, testing, and standardization.
- Incentive policies that are not consistent: The lack of consistent, financially appealing incentive schemes in various markets breeds uncertainty and impedes the expansion of small wind energy systems.

A wind turbine's power production is determined by a number of variables, such as wind speed, turbine size, and design. The power generated is stated as follows:

$$P = ½ CpUV*3 A$$

where A is the swept area of the rotor blade in square meters (m^2), V is the wind speed in meters per second (m/s), U is the air density in kilograms per cubic meter (kg/m^3), P is the power in watts (W), and (C_p) is the power coefficient.

Rating systems, testing protocols, and performance standards are crucial. The implementation of state rebate schemes for small wind installations has been hampered by the lack of uniform ratings and clear requirements. By establishing these guidelines, the industry's credibility would be enhanced, and consumers would receive trustworthy information for well-informed purchasing decisions. Enhancing wind turbine efficiency, dependability, and durability will require significant technological advancements. For wind energy to be economically viable in regions with low wind speeds, technological advancements in turbines are essential. Improvements in electronics and integration systems, as well as supervisory control systems for optimizing turbine operations in conjunction with load management, should be the main areas of research and development.

10.5 SOLAR PV

One of the most popular renewable energy sources is solar energy, which is completely free, infinite, and emission-free. It can be used to produce heat and power. Numerous PV technologies have been created, tested, and released onto the market in recent years. Every technology has unique features in terms of dependability, efficiency, and design. With a lifespan of more than 25 years, PV technology has grown more dependable as it has developed. Three types of PV generators can be distinguished.

Crystalline silicon (Si) cells, which account for over 87% of the market and have an average efficiency of 20%, make up the first group of PV cells. Thin-film solar cells of the second kind include cadmium telluride (CdTe), copper indium gallium selenide (CIGS), and amorphous silicon. CdTe is the most practical of these materials because it strikes a compromise between cost and efficiency. PV cells with a high efficiency of about 40% make up the third category. The III-V multi-junction solar cells (GaInP/GaInAs/Ge), which are regarded as the third generation of PV, now possess the most sophisticated technology. These cells work quite well, but they have a significant disadvantage: they are very expensive, which restricts their use to concentrated photovoltaic (CPV) systems. Very tiny solar cells are used in CPV modules, which lowers overall prices and material consumption.

A DC–AC converter (inverter), which also helps the PV system perform its maximum power point tracking (MPPT) function, can be used to feed the generated energy into the grid. As an alternative, MPPT can be used at the level of PV generators. A distributed MPPT DC–DC converter is utilized in this instance to guarantee that every PV module runs at its ideal power point. This method improves the overall performance and efficiency of the system (Figure 10.2).

10.5.1 Distributed PV generation obstacles

Even with the extensive installation of PV plants, a number of market-related barriers still prevent dispersed PV applications from growing:

Cost inequality: PV systems continue to produce electricity at a higher cost than conventional fossil fuel sources. Customers can find it difficult to understand the significance of backing renewable energy projects as long as conventional energy is still less expensive.

High initial investment and long payback period: Despite a slow decline in PV system costs, there are still substantial expenditures associated with PV devices, equipment, and plant design. As a result, the time needed to recoup the initial investment is longer than the PV plant's normal lifespan.

Complicated financing procedures: People are deterred from pursuing financial assistance for PV installations by drawn-out bureaucratic procedures and intricate financing procedures.

Growing distributed PV generation is difficult: Modern PV systems are challenging to integrate into electricity markets that are primarily structured around centralized power plants.

Inadequate government support: To promote the use of PV, Italy launched the "Conto Energia" incentive program in 2005, which attracted foreign

Figure 10.2 Solar PV.

investors until 2011. The updated 2012 regulation, however, made it harder to obtain incentives, which hindered the expansion of the PV market and had detrimental effects on industry participants.

In addition to market-related difficulties, technical barriers must also be taken into account. The visual impact of PV plants is one significant problem. Installing PV systems in locations with stringent landscape preservation regulations is still difficult, even with the availability of numerous building integrated PV (BIPV) solutions and technology.

10.5.2 Prospects for the future

To create smarter PV plants, methods and approaches will keep changing in the upcoming years. The performance of these cutting-edge systems will be better than that of current PV configurations. Higher efficiency technologies will be incorporated to accomplish this, and they will also adjust to shifting operating conditions and shading problems. Dynamic techniques that adjust PV plant layouts as necessary (matrix reconfiguration) will be used to accomplish this adaptability.

Furthermore, since smart grids are expected to expand, PV systems will communicate with DG operators and other network elements more frequently. To guarantee the controlled and optimal administration of the electrical grid, this interaction will contribute to the provision of useful ancillary services and data. The development of appropriate technology and the establishment of appropriate grid codes are necessary to facilitate this shift.

10.6 FUEL CELL

10.6.1 Technology for FCs

An FC is a machine that uses electrochemical processes to transform chemical energy into heat and electrical energy. As long as it has a steady supply of fuel, it works similarly to a battery, producing electrical energy. An FC does not need to be recharged like conventional batteries do since the materials used in the electrochemical process are constantly being replaced. Without using combustion, FC technology produces electricity through an electrochemical reaction between hydrogen and oxygen. When compared to traditional generators, this process generates direct current (DC) power, water, heat, and very little emissions (such NOx and CO_2), making it an environmentally benign way to generate electricity. Because of their ability to achieve very high electrical efficiency, FCs are regarded as a possible energy source for the future. Based on the kind of electrolyte they utilize and their working temperature, FCs are classified into different varieties and use different technologies. The direct methanol fuel cell (DMFC), phosphoric acid fuel cell (PAFC), alkaline fuel cell (AFC), and proton exchange membrane fuel cell (PEMFC) are a few types of low-temperature FCs. Conversely, solid oxide fuel cells (SOFCs) and molten carbonate fuel cells (MCFCs) are examples of high-temperature FCs. Proton exchange membrane fuel cell, or PEMFC: utilized in both residential and commercial settings for cogeneration, or the combination of heat and power. Phosphoric acid fuel cell, or PAFC: Cogeneration applications are appropriate for light, heavy, and commercial industries. Molten carbonate fuel cell, or MCFC: utilized in grid-sited power production plants (distributed and centralized). utilized in heavy, light, and commercial contexts for cogeneration as well. Solid oxide fuel cell, or SOFC: Grid - sited applications (both centralized and distributed power generating) use it similarly to MCFC. Additionally, it can be used for cogeneration in a variety of industries, such as light, heavy, commercial, and residential.

10.6.2 FCs for distributed and power generation

FCs can be used in a variety of ways, from small units (about 1 kW) meant for backup power to multi-megawatt systems intended for big centralized power generation. They may power buildings, houses, and DG systems, including CHP applications, because they come in sizes ranging from 1 kW to 3 MW. By providing a range of services, FCs provide flexible energy solutions in the context of DG. Their main energy source is hydrogen, which needs to be collected from chemical sources like water, natural gas, or other carriers. Using renewable energy is the most ecologically friendly way to produce hydrogen. Hydrogen can be extracted from water using a process called electrolysis, which can be powered by solar or wind energy. Nevertheless, there are still certain energy losses and extra expenses associated with turning renewable electricity into hydrogen and back into electrical energy. Compared to conventional fossil fuel-based systems, FCs can achieve an approximately 100% decrease in emissions when fueled by renewable hydrogen.

10.6.3 FC benefits for DG

- Reduction of distribution and transmission losses
- Minimal emissions that help create a cleaner environment
- Increased power supply dependability
- Less stress on the electrical grid, especially when demand is at its highest
- Operation that is silent and vibration-free, reducing noise pollution
- Reduced requirements for upkeep and operation
- Outstanding load-following and transient response capabilities

FCs can maximize the use of fuel energy in CHP applications by achieving above 80% efficiency. Furthermore, FCs can be powered by fuels sourced from biomass, which has the added benefit of reducing greenhouse gas emissions and making better use of biomass resources.

10.6.4 FC integration challenges in DG

Cost and durability are the two biggest barriers to the development of clean, dependable, and affordable FC systems, while there are other important issues as well. Even though FCs have become smaller and lighter, more innovations are still needed to satisfy the packing requirements of some commercial applications. It is crucial to gain a better knowledge of how air, fuel, and system impurities—including those from fuel storage systems—affect FC efficiency. Strategies to lessen these effects also need to be determined and clearly illustrated. There are still significant obstacles to the broad commercialization of FCs, including the cost, effectiveness, and packaging of balance-of-plant (BOP) components, or the auxiliary equipment needed for a FC system. Improving FC performance is essential for increasing cogeneration capacities and system efficiency in stationary power applications. To promote the use of renewable energy sources like biogas and allow for more fuel flexibility, it is imperative to develop low-cost technology for gas clean-up and fuel processing. Furthermore, enhancing the long-term dependability and commercial success of high-temperature FC systems, such as MCFC and SOFC, requires lowering their cost while increasing their durability.

1. The main obstacle is durability: Enhancing durability is a major difficulty in FC development. Stationary FCs need to operate for more than 60,000 hours to effectively compete with alternative distributed power-generating systems and guarantee a fair return on investment for end users. Because of their high operating temperatures, high-temperature FCs are subject to extremely demanding conditions that put a great deal of strain on system materials and components, including the electrolyte, electrolyte support, and electrodes. Enhancing the durability of BOP components is essential, regardless of the particular application. Notably, non-stack BOP components, not the FC stack itself, are responsible for about 90% of failures and unplanned outages in micro-CHP systems.
2. Performance and costs: Significant cost reductions are required for FCs and accompanying systems to become commercially viable. It is crucial to reduce the materials and production costs related to FC stack components. Creating a cost-effective procedure and subsystem for efficiently eliminating impurities, especially those found in renewable fuels, is a significant issue. Resolving this issue would improve fuel flexibility in addition to lowering total expenses. Certain BOP elements, including heat exchangers, need to be able to function at high temperatures in order to guarantee peak performance for high-temperature FCs. FCs and their systems must perform at efficiency levels that are on par with or higher than those of rival technologies. Reaching this goal is essential to increasing market uptake and optimizing FC technology's environmental advantages.

10.6.5 FCs' prospects

There are several possible uses for stationary FCs in the commercial, industrial, and residential domains. They can be used in addition to or instead of conventional electrical grid power sources. In comparison to traditional systems, FCs can improve energy efficiency, lower emissions, and provide more reliability in this way. Power outputs from FCs are anticipated to range from 25 kW to 500 kW in the small business sector. This function can be performed by a variety of FC types, serving facilities including hotels, shopping malls, schools, hospitals, and office buildings. Cogeneration mode is anticipated to be used by high-temperature FCs, such as MCFCs and SOFCs, to provide both heat and electricity at the same time. FCs will compete in this market with other cutting-edge technologies like microturbines in addition to one another. FCs are expected to produce power outputs in the industrial sector that range from 1 MW to 25 MW. High-temperature FCs, which are ideal for sectors like plastics, food production, metal processing, paper manufacture, and chemical production, will be a major component of this market. FCs will be able to provide electricity in the 3 MW to 100 MW range for the distributed power industry. Energy service companies, unregulated subsidiaries, municipal utilities, and traditional utilities will all be a part of this market. FCs will produce modest power outputs in the home market, usually ranging from 1 kW to 10 kW. This industry is anticipated to be served by PEMFCs and SOFCs, which will first power single-family and multi-family homes in electric-only arrangements. Data centers and critical infrastructure are examples of specialized businesses that FCs may be able to service since they demand high reliability and premium power quality. Furthermore, biogas (from waste or wastewater treatment plants), landfill gas, and fuels from biorefineries are examples of renewable fuels that could be used with FCs. Combining high-temperature FCs with fuel gas made from alternative energy sources like biomass, sewage sludge, manure, and organic waste from the food and agriculture industries offers a promising way to generate energy sustainably while maximizing energy efficiency and reducing environmental impact.

10.7 CONCLUSION

Given the wide range of potential uses for DG and the growing trend of many developed countries switching to renewable energy sources, DG may be a major factor in power generation in the future. However, in order to overcome the current obstacles that DG systems face, more research and development will be necessary to realize this potential. DG technologies need to concentrate on lowering capital costs, limiting emissions, and increasing efficiency in order to prosper. The importance of DG systems should also be acknowledged by government policymakers and electricity market regulators. Market structures, operational frameworks, and pricing models can be modified to facilitate the expansion and sustained prosperity of DG technologies.

REFERENCES

1. Twaisan, K., & Barışçı, N. (2022). Integrated distributed energy resources (DER) and microgrids: Modeling and optimization of DERs. *Electronics*, *11*(18), 2816.
2. Conti, J., Holtberg, P., Diefenderfer, J., LaRose, A., Turnure, J. T., & Westfall, L. (2016). *International energy outlook 2016 with projections to 2040* (No. DOE/EIA-0484 (2016)). USDOE Energy Information

Administration (EIA), Washington, DC (United States). Office of Energy Analysis.

3. Newell, R., Raimi, D., & Aldana, G. (2019). Global energy outlook 2019: The next generation of energy. *Resources for the Future*, *1*(8), 1–37.
4. Mishra, P., & Singh, G. (2023). Energy management systems in sustainable smart cities based on the internet of energy: A technical review. *Energies*, *16*(19), 6903.
5. Bhatti, H. J., & Danilovic, M. (2018). Making the world more sustainable: Enabling localized energy generation and distribution on decentralized smart grid systems. *World Journal of Engineering and Technology*, *6*(2), 350–382.
6. Alotaibi, I., Abido, M. A., Khalid, M., & Savkin, A. V. (2020). A comprehensive review of recent advances in smart grids: A sustainable future with renewable energy resources. *Energies*, *13*(23), 6269.
7. National Academies of Sciences, Medicine, Division on Engineering, Physical Sciences, Board on Energy, Environmental Systems, ... & Distribution System. (2017). *Enhancing the resilience of the nation's electricity system*. National Academies Press.
8. Rehmani, M. H., Reisslein, M., Rachedi, A., Erol-Kantarci, M., & Radenkovic, M. (2018). Integrating renewable energy resources into the smart grid: Recent developments in information and communication technologies. *IEEE Transactions on Industrial Informatics*, *14*(7), 2814–2825.
9. Twaisan, K., & Barışçı, N. (2022). Integrated distributed energy resources (DER) and microgrids: Modeling and optimization of DERs. *Electronics*, *11*(18), 2816.
10. Anusha, G., Arora, K., Sharma, H., Thota, S. P., Joshi, G. P., & Cho, W. (2024). Control strategies of 15-level modified cascaded H-bridge MLI with solar PV and energy storage system. *Energy Reports, 12*, 2–26. https://doi.org/10.1016/j.egyr.2024.06.003
11. Lopes, J. P., Hatziargyriou, N., Mutale, J., Djapic, P., & Jenkins, N. (2007). Integrating distributed generation into electric power systems: A review of drivers, challenges and opportunities. *Electric Power Systems Research*, *77*(9), 1189–1203.
12. Allan, G., Eromenko, I., Gilmartin, M., Kockar, I., & McGregor, P. (2015). The economics of distributed energy generation: A literature review. *Renewable and Sustainable Energy Reviews*, *42*, 543–556.
13. Kyriakopoulos, G. L., & Arabatzis, G. (2016). Electrical energy storage systems in electricity generation: Energy policies, innovative technologies, and regulatory regimes. *Renewable and Sustainable Energy Reviews*, *56*, 1044–1067.
14. Pehnt, M., Cames, M., Fischer, C., Praetorius, B., Schneider, L., Schumacher, K., & Voß, J. P. (2006). *Micro cogeneration: towards decentralized energy systems*. Springer Science & Business Media.
15. Jiménez, M. S. (2007). *Smart electricity networks based on large integration of renewable sources and distributed generation* (Vol. 3). Kassel University Press GmbH.

16. Pepermans, G., Driesen, J., Haeseldonckx, D., Belmans, R., & D'haeseleer, W. (2005). Distributed generation: Definition, benefits and issues. *Energy Policy*, *33*(6), 787–798.
17. Blaabjerg, F., Chen, Z., & Kjaer, S. B. (2004). Power electronics as efficient interface in dispersed power generation systems. *IEEE Transactions on Power Electronics*, *19*(5), 1184–1194.
18. Lopes, J. P., Hatziargyriou, N., Mutale, J., Djapic, P., & Jenkins, N. (2007). Integrating distributed generation into electric power systems: A review of drivers, challenges and opportunities. *Electric Power Systems Research*, *77*(9), 1189–1203.
19. Gallegos, J., Arévalo, P., Montaleza, C., & Jurado, F. (2024). Sustainable electrification—Advances and challenges in electrical-distribution networks: A review. *Sustainability*, *16*(2), 698.
20. Bajaj, M., & Singh, A. K. (2020). Grid integrated renewable DG systems: A review of power quality challenges and state-of-the-art mitigation techniques. *International Journal of Energy Research*, *44*(1), 26–69.
21. Abdelkader, S., Amissah, J., & Abdel-Rahim, O. (2024). Virtual power plants: An in-depth analysis of their advancements and importance as crucial players in modern power systems. *Energy, Sustainability and Society*, *14*(1), 52.
22. Heichel, G. H. (1976). Agricultural Production and Energy Resources: Current farming practices depend on large expenditures of fossil fuels. How efficiently is this energy used, and will we be able to improve the return on investment in the future?. *American Scientist*, *64*(1), 64–72.

Chapter 11

Renewable energy integration and grid interfacing

Akanksha Sharma

11.1 INTRODUCTION

The shift to renewable energy sources (RES) is central to addressing global energy issues, including climate change, energy security, and environmental sustainability. The countries are focusing on reducing carbon emissions and decreasing dependence on fossil fuels. Preferences are being given to viable RES alternatives, namely solar, wind, and hydro. However, despite substantial developments in renewable energy (RE) technologies, their intermittent nature presents significant challenges in ensuring a reliable energy supply, maintaining grid stability, and managing resources efficiently.

Smart grid technologies are being used globally in response to these challenges, as they offer a more flexible, responsive, and resilient energy infrastructure. The integration of RES is not only transforming energy generation but also accelerating the evolution of smart grids by necessitating advanced control, communication, and storage solutions [1]. These developments support the dual goals of improving grid performance and enabling the large-scale use of clean energy. These challenges are predominantly evident in systems that depend heavily on weather-dependent resources like solar and wind energy, where energy production can vary unpredictably based on weather conditions and seasonal variations.

The RES incorporation within smart grid has arisen as a key approach to mitigate these issues. Smart grids are equipped with advanced technologies that help in real-time monitoring, data analytics, and adaptive control mechanisms, permitting better integration of distributed RE generation. Smart grids decrease dependency on centralized power plants and increase the grid's overall stability and efficiency by making the power system more responsive and resilient. [2]. Intelligent control and real-time communication enable dynamic demand and supply balancing, maximize the energy use from intermittent sources, and maintain operational reliability [3].

Additionally, the declining cost of RE technologies, particularly photovoltaic (PV) systems, has accelerated their deployment across residential, commercial, and utility-scale sectors. PV systems are among the most widely

DOI: 10.1201/9781003689430-11

adopted RES due to their scalability and environmental benefits [4]. However, their performance depends significantly on constant monitoring and maintenance to ensure optimal generation efficiency over time. By storing excess energy during times of high availability and supplying it during times of low production, energy storage systems (ESS) can also work together with RES. ESS is a crucial part of smart grid-enabled RE integration owing to its ability to increase system flexibility, lesser RE curtailment, and a more dependable energy supply. It can be said from a broader viewpoint that the integration of RES within smart grids plays a critical role in advancing the United Nations sustainable development goals (SDGs), specifically SDG 7 (Affordable and Clean Energy) and SDG 13 (Climate Action). By enabling cleaner energy transitions and reducing greenhouse gas emissions, this synergy supports global efforts toward environmental sustainability and equitable energy access [5].

Despite the growing adoption of RES owing to significant environmental and economical benefits, their integration into the existing electricity grid introduces several technical and operational challenges. The most prominent among these is their intermittent and variable nature, which can lead to imbalances in supply and demand, voltage fluctuations, and frequency instability. Solar and wind energy, for instance, are weather-dependent and can produce power unpredictably, making it difficult to ensure steady energy availability and system reliability.

One of the key concerns is the mismatch between generation and demand due to the weather and season dependence of many RES [6]. For example, PV generation peaks during daytime, which may not coincide with peak consumption time always. Wind energy also fluctuates based on weather conditions, which further complicates the load balancing and necessitates flexible system operations. Additionally, the issue also lies in the lack of inertia and synchronous characteristics in inverter-based RE systems. Traditional power plants provide system inertia through rotating masses, which help in stabilizing the frequency during disturbances. In contrast, most RES units connected via power electronic interfaces now do not inherently contribute to inertia. Thus, the reduction in system inertia can destabilize grid operations, making it difficult for grid-following converters to maintain voltage and frequency stability [7]. Therefore, grid interfacing technologies become crucial. Grid interfacing refers to the combination of hardware and software components that facilitate the smooth connection, control, and coordination between RES and the main grid. Power electronic converters, smart inverters, and advanced control systems play a central role in this process. These interfaces enable voltage and frequency regulation, reactive power management, islanding detection, and grid synchronization—all vital for the stable operation of a renewable-rich power system.

Moreover, grid interfacing supports bidirectional power flow, which is essential in decentralized energy systems where consumers can also act as producers and therefore known as prosumers [8]. It also helps in implementing grid codes, supporting fault ride-through capabilities, and ensuring compatibility across different voltage levels and system configurations.

As smart grid infrastructure is evolving, the role of grid interfacing is expanding beyond basic connection functionalities. Modern interfacing systems are increasingly equipped with real-time data acquisition, artificial intelligence (AI)-based control algorithms, and self-healing capabilities, thereby strengthening grid resilience and flexibility.

This chapter provides a thorough understanding of the integration of RES with power grids. It starts with the overview of RES, namely geothermal, biomass, hydro, wind, and solar, emphasizing their key characteristics, challenges, and opportunities available. The chapter then discusses managing the variability of these sources. The chapter examines smart grid solutions that allow effective integration, including demand response (DR), energy storage, advanced control systems, and power electronics interfaces. Afterwards, real-world case studies and industry applications are presented to show practical implementations and lessons learnt. The chapter ends with the main findings and future directions for improving RE integration through smart grid strategies.

11.2 RE SYSTEMS BASICS

Unlike traditional fossil fuels, RES such as solar power, wind, hydropower, biomass, geothermal, and tidal energy are abundant, naturally replenishing, and environmentally friendly. These systems offer a sustainable alternative to diminishing fossil fuels, while also helping to lower greenhouse gas emissions, combat climate change, and enhance energy security [9, 10].

Figure 11.1 highlights the major RES, each using distinct technologies to harness natural resources for electricity generation. Understanding the basic principles behind these technologies is important for designing efficient, reliable, and economically viable RE systems. Figure 11.2 shows the growth of global RE capacity from 2015 to 2024, highlighting substantial expansion across all major sources, with solar and wind undergoing the most rapid increases. Hydropower remains the largest contributor in terms of capacity, while biomass, geothermal, and marine energy have shown comparatively slower growth [11].

11.2.1 Solar power

Solar energy is a renewable resource that harnesses the sun's energy by converting it into electricity or heat. It is the most abundant and sustainable energy source, which plays a crucial role in shifting toward a cleaner and eco-friendly global energy future.

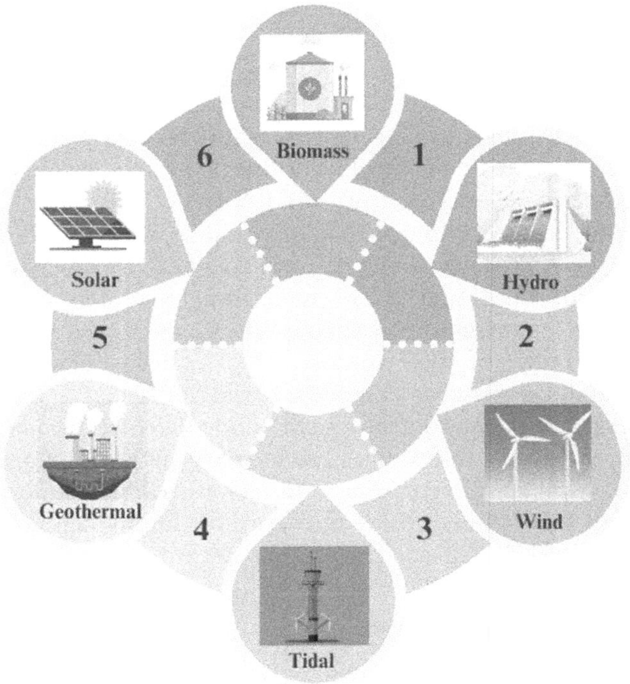

Figure 11.1 Main sources of renewable energy.

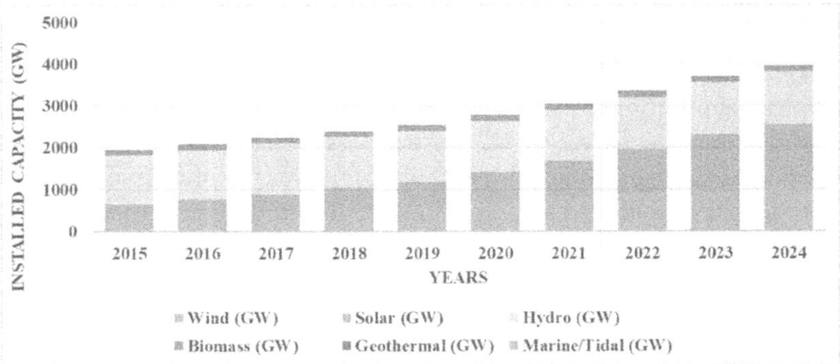

Figure 11.2 Global installed capacity by source from 2015 to 2024.

11.2.1.1 Characteristics of solar power

- **Abundant and renewable:** The sun provides a limitless amount of energy, thus making solar power an inexhaustible resource.
- **Zero emissions:** Solar energy generation has no greenhouse gases or harmful pollutants, thus making it an eco-friendly option.
- **Scalable and flexible:** Solar power can be utilized at diverse levels, ranging from small rooftop installations to large-scale solar power plants.

11.2.1.2 Challenges and opportunities

- **Intermittency:** Solar energy generation depends on sunlight, which can fluctuate due to changes in weather conditions and the position of the sun throughout the day.
- **Energy storage:** Effective storage solutions, like batteries, are essential to retain SE for use during times when sunlight is not present
- **Initial investment:** The initial expense of buying and setting up solar panels can be substantial, although costs have significantly decreased in recent years.
- **Space considerations:** Large solar farms need considerable amounts of land, though rooftop installations minimize this issue.

11.2.2 Wind power

It is a clean RES that transforms the kinetic energy of moving air into electricity. Wind power is crucial in lowering greenhouse gas emissions and supporting the global shift toward sustainable energy systems.

11.2.2.1 Key characteristics of wind power

- Renewable and abundant: Wind energy is freely available in many regions, mainly in coastal and open plains areas.
- Low environmental impact: Wind turbines produce electricity without producing greenhouse gases, thus are environmentally friendly.
- Scalable solutions: Wind power systems range from small turbines for individual or community use to large-scale wind farms.

11.2.2.2 Challenges and opportunities

- **Intermittent nature:** Wind energy depends on wind availability; thus, it is not constant energy sources and varies according to weather conditions, making it variable and unpredictable.

- **Land use and aesthetics:** Wind farms require large areas, and their visual impact can lead to community concerns.
- **Wildlife concerns:** Wind turbine blades may affect bird population, necessitating environmentally friendly designs.
- Advancements in turbine efficiency, materials, and energy storage are improving the reliability and affordability of wind power. Emerging technologies like floating offshore wind turbines and advanced ESS, along with innovations in grid integration and hybrid renewable systems, are expanding wind energy's potential and adaptability in modern energy systems.

11.2.3 Hydropower

It is also known as hydroelectric power, which captures the energy from flowing water to generate electricity. This long-established and widely adopted RES is essential in providing clean, reliable, and efficient power to communities globally.

11.2.3.1 Characteristics of hydropower

- **Renewable resource:** Water is replenished by the natural water cycle, thus providing a sustainable and abundant source of energy.
- **Reliable and consistent:** Hydropower can generate continuous energy production, particularly from reservoirs with controlled water flow.
- **Scalable:** Hydropower systems range from large-scale dams to small, localized micro-hydro setups, accommodating diverse energy needs.

11.2.3.2 Challenges and opportunities

- **Environmental impacts:** Dams and reservoirs have the potential to impact ecosystems, hinder fish migration, and affect water conditions.
- **Dependency on water availability:** Seasonal changes and droughts can affect water flow, impacting energy production.
- **High initial investment:** Constructing a large hydropower plant needs significant financial resources and infrastructure. Thus, the initial cost is high.
- **Technological innovations:** Advances in turbine design, fish-friendly technologies, and small hydro systems are addressing many of these challenges.

11.2.4 Biomass energy

This energy emanates from organic materials like plants, agricultural wastes, forest by-products, and other biological waste. By transforming these materials into thermal energy, electricity, or biofuels, biomass energy helps in reducing the dependence on fossil fuels and encourages sustainable energy solutions.

11.2.4.1 Characteristics of biomass energy

- **Renewable resource:** Biomass is naturally replenished by the growth of plants and trees, making it a sustainable energy source.
- **Carbon neutrality:** Biomass energy has the potential to be carbon neutral if handled properly, as plants absorb carbon dioxide while growing, thus, can balance the emissions produced during their conversion into energy.
- **Diverse feedstocks:** A wide range of organic materials can be used, including crop residues, wood, manure, and municipal waste.

11.2.4.2 Challenges and opportunities

- **Land and resource use:** Growing biomass material requires land availability and careful land-use planning.
- **Emission concerns:** Although biomass is often carbon neutral, incomplete combustion or unsustainable practices can result in greenhouse gas emissions.
- **Efficiency improvements:** Developing advanced technologies for biomass conversion can improve energy output and lessen the environmental impact.
- **Sustainable practices:** Effective policies and management are essential to ensure the sustainable production and utilization of biomass.

11.2.5 Geothermal energy

It harnesses the Earth's natural heat as a sustainable source of energy. This energy can be effectively tapped for producing electricity, providing heating, and powering various applications.

11.2.5.1 Characteristics of geothermal energy

- **Continuous supply:** Unlike wind or solar, geothermal energy is available continuously, providing a consistent and reliable power source.
- **Environmentally friendly:** Geothermal systems emit minimal greenhouse gases compared to fossil fuels, making them a clean energy option.
- **Scalability:** It can support both large-scale power plants and localized heating systems, catering to diverse energy needs.

11.2.5.2 *Challenges and opportunities*

- **High initial costs:** Drilling and exploration require substantial investment, though operating costs are low once systems are established.
- **Geographical limitations:** Geothermal resources are concentrated in certain regions, limiting widespread accessibility.
- **Environmental concerns:** Challenges like water consumption, land sinking, and minor gas emissions need careful management.
- **Technological advancements:** Inventions in enhanced geothermal systems are expanding the feasibility of accessing deeper and less accessible geothermal resources.

11.2.6 Ocean energy

It involves various forms of RE generated by the natural movement and temperature of seawater. This resource encompasses energy from tides, waves, ocean temperature differences, and salinity gradients. Ocean energy holds significant promise for driving the global transition to clean and sustainable energy solutions.

11.2.6.1 *Characteristics of ocean energy*

- **Sustainable and predictable:** Ocean energy sources, such as tidal movements and waves, follow consistent natural cycles, providing reliable energy generation.
- **Diverse energy options:** The ocean offers multiple mechanisms for energy extraction, ensuring adaptability for different needs.
- **Eco-friendly:** These systems produce little to no greenhouse gas emissions during their operation.

11.2.6.2 *Types of ocean energy*

- **Tidal energy:** This energy is produced by the gravitational pull of the moon and sun, which causes the movement of tides. Technologies like tidal barrages, tidal stream generators, and tidal lagoons are used to capture this energy.
- **Wave energy:** Wave energy harnesses the power of surface waves using devices such as oscillating water columns, point absorbers, and wave energy converters.
- **Ocean thermal energy conversion:** This method uses the temperature difference between warm surface waters and cooler deep waters to generate electricity.

11.2.6.3 Benefits of ocean energy

- **Abundant eesource:** Oceans cover over 70% of the Earth's surface, providing an enormous RE resource.
- **Reliable output:** Tidal and current energy are highly predictable compared to other renewable sources like solar and wind.
- **Localized energy production:** Coastal areas can utilize nearby ocean energy resources, reducing dependence on external energy supplies.

11.2.6.4 Challenges and opportunities

- **Technological barriers:** The technology for ocean energy extraction is still developing and requires further refinement to become commercially viable.
- **Environmental considerations:** It is essential to minimize potential impacts on marine ecosystems during the deployment of ocean energy systems.
- **Economic hurdles:** The high initial investment and maintenance costs of these systems are obstacles to widespread adoption.
- **Infrastructure needs:** Expanding coastal infrastructure for energy extraction and transmission is vital for scaling up ocean energy utilization.

11.2.7 Fuel cells and their renewable potential

It is a device that converts chemical energy from a fuel into electrical energy using an electrochemical process. Noted for its efficiency and eco-friendly nature, it shows great potential as a RES, especially when using hydrogen derived from sustainable processes.

11.2.7.1 Characteristics of fuel cells

- **Enhanced efficiency:** Fuel cells bypass intermediate thermal processes, converting chemical energy directly into electrical energy, making them more efficient than traditional combustion-based systems.
- **Eco-friendly operation:** When fuelled with hydrogen, the primary by-products are water and heat, classifying fuel cells as a zero-emission technology.
- **Wide applicability:** Fuel cells serve various sectors, including transportation (like hydrogen-powered vehicles), stationary energy systems, and portable power solutions.
- **Hydrogen as a RES**

Fuel cells can achieve true sustainability when powered by green hydrogen, which is produced using renewable methods such as:

- **Electrolysis of water:** Utilizing RE like solar or wind to split water into hydrogen and oxygen.
- **Biomass gasification:** Generating hydrogen-rich gas from organic materials.
- **Photobiological and photoelectrochemical techniques:** Using sunlight and biological systems or catalysts to produce hydrogen.

11.2.7.2 Benefits of fuel cells in RE systems

- **Energy storage:** Fuel cells can store energy by producing hydrogen, which can later be converted back to electricity, supporting intermittent renewable sources like solar and wind.
- **Localized power generation:** They facilitate decentralized energy production, minimizing transmission losses.
- **Smart grid integration:** Fuel cells enhance the reliability and flexibility of smart grids and microgrids.

Challenges and future prospects

- **Cost reduction:** The high initial costs and reliance on expensive materials like platinum for catalysts remain significant hurdles, prompting ongoing research into affordable alternatives.
- **Hydrogen infrastructure:** A comprehensive system for hydrogen production, storage, and distribution is critical for widespread adoption.
- **Technological advancements:** Advancements in hydrogen production and renewable integration will enhance the contribution of fuel cells to sustainable energy systems.

11.2.8 Summary

This section provides an in-depth overview of different RE systems, such as solar, wind, hydro, geothermal, ocean, biomass, and fuel cells. It delves into the fundamental characteristics and different types of each energy source, providing a clear understanding of their unique attributes. Additionally, the section addresses the challenges these systems face, such as intermittency, resource availability, and technological limitations. By investigating these issues in depth, the section highlights the potential of RES in shaping a cleaner and more resilient energy landscape.

11.3 MANAGING THE VARIABILITY OF RES

RES like wind, solar, and hydropower play a crucial role in building a sustainable, low-carbon energy system. However, their variable nature presents challenges in maintaining a consistent and reliable energy supply, as availability fluctuates with weather patterns, time periods throughout the day, and seasonal changes. Managing these fluctuations is crucial for grid stability and power reliability. Addressing RES variability requires innovative approaches, technological advancements, and strategic planning to enhance their incorporation into the power grid while ensuring energy security and reliability. The following strategies outline key methods for managing this variability.

11.3.1 Energy storage solutions

- **Batteries:** Batteries, including advanced types like lithium-ion, sodium-nickel chloride, and flow batteries, play a crucial role in addressing the variability of RES. These technologies store excess energy during high production periods, such as peak solar or wind output, and release it during low generation times to balance demand.
  Lithium-ion batteries, known for their high energy density, rapid charge/discharge rates, and long lifespan, are ideal for applications requiring quick response, such as grid-scale or distributed systems. Sodium-nickel chloride batteries, offering high efficiency, moderate operating temperatures, and non-toxic materials, are suited for medium to long-duration storage, aiding grid stability in renewable-heavy systems [12]. Flow batteries, with their liquid electrolyte design, excel in large-scale, long-duration storage, effectively addressing RE fluctuations. These battery systems also support grid services like frequency regulation, ensuring a stable and reliable power supply.
- **Pumped hydro storage:** Pumped hydro storage (PHS) is a proven and efficient method for large-scale energy storage, utilizing two reservoirs at different elevations. Excess electricity from renewable sources, such as wind and solar, powers pumps that transfer water from a lower to an upper reservoir, storing energy as gravitational potential. During high demand or low renewable output, the stored water is released to generate electricity through turbines.
  With an efficiency of 70–85%, PHS is highly effective for stabilizing grids and ensuring a steady power supply. Its long lifespan makes it a reliable, long-term storage solution, particularly valuable for managing RE variability in regions with suitable topography and water resources. By storing surplus energy during peak generation and releasing it during deficits, PHS helps maintain grid stability as RE penetration grows [13].

- **Thermal storage:** Thermal energy storage (TES) systems, often using molten salts, store heat from renewable sources like solar and wind. During peak generation, molten salts are heated and stored, producing steam to drive turbines when demand rises or generation drops, stabilizing grids as variable RE increases. TES also enhances district heating systems by improving heat-load flexibility and reducing power surpluses and deficits by about 9%, cutting fuel consumption by 10%, supporting grid stability [14].

  Advancements include phase change materials that store energy during solid–liquid transitions, with rib-enhanced plates improving melting times and efficiency [15]. Solar-driven TES with absorption heat pumps enhances flexibility and efficiency, addressing renewable intermittency [16]. Emerging electric and hydrogen TES technologies offer scalable, cost-efficient seasonal storage solutions, particularly for smart cities like Neom, relying on wind and solar energy [17]. Combining TES with wind, bio-waste, and hydrogen in flexi-renewable hubs optimizes energy use, improving efficiency by up to 24% [18].
- **Hydrogen storage:** Hydrogen storage is a vital solution for managing the variability of RES like solar and wind. Excess electricity is converted into hydrogen through electrolysis, splitting water into hydrogen and oxygen. The hydrogen is stored as compressed gas, liquid hydrogen, or in chemical forms like ammonia or metal hydrides, enabling long-term energy retention from days to months. Unlike traditional batteries, hydrogen storage bridges gaps caused by weather, time of day, or seasons, stabilizing power grids and enhancing reliability. With its high energy density, hydrogen storage supports large-scale applications in power generation, industrial processes, and transportation. It improves grid flexibility by storing surplus energy during low demand and utilizing it during high demand or low renewable output, reducing curtailment and increasing efficiency [19].

  To integrate RES like solar and wind effectively, various energy storage technologies, including batteries, PHS, hydrogen storage, and TES, are used. TES stands out as a cost-effective option for storing large amounts of energy, particularly in conjunction with solar power [20]. However, hydrogen storage is critical for mitigating RES variability, optimizing system performance, and ensuring reliability. Combining hydrogen and thermal storage offers a robust, flexible solution for meeting both short-term and long-term energy storage demands.

11.3.2 Grid enhancement with smart technology

- **Advanced grid infrastructure:** Smart grids are a significant advancement in energy distribution systems, designed to efficiently manage fluctuating inputs from RES like wind and solar. Smart grids use sensors, data analytics, and cutting-edge technologies to optimize energy distribution, cut waste, and improve grid reliability by responding to supply and demand changes in real time. They support ESS and electric vehicles (EVs), improving the energy infrastructure's flexibility and resilience. This automation makes smart grids indispensable for a sustainable energy future.

- **DR**: It addresses the variability of RES by incentivizing consumers to adjust electricity usage during periods of low renewable generation or high demand. For instance, users may shift consumption to times of abundant RE, such as daytime solar peaks, or reduce usage during peak demand. This reduces energy costs, optimizes grid efficiency, and minimizes reliance on fossil fuel-based backup power [21]. DR enhances grid resilience, balancing intermittent renewable output and supporting the integration of higher RES shares, ensuring a stable and sustainable energy system.

11.3.3 Hybrid systems for enhanced stability

- **Wind-solar hybrid systems:** Hybrid wind-solar systems leverage the complementary nature of these resources, with solar power peaking during the day and wind energy often strongest at night or in low-sunlight periods. This balance reduces power fluctuations, enhancing grid stability and ensuring a more reliable energy supply, particularly in regions where neither resource alone can consistently meet demand.
- **Integration with conventional sources:** Pairing RES with conventional power plants like natural gas, coal, or nuclear ensures a stable power supply during low RE generation periods, such as cloudy weather or low winds. This hybrid approach combines the flexibility of renewables with the reliability of conventional plants, stabilizing the grid and preventing disruptions while adapting to fluctuations in generation and demand.

11.3.4 Flexible and rapid-response generation

- **Gas turbines and combined cycle plants:** Gas turbines, especially in combined cycle plants, provide a flexible solution for balancing RE fluctuations. These plants can quickly adjust output to compensate for variability in wind or solar generation, ensuring a reliable backup when renewable resources are insufficient. Combined cycle systems, utilizing both gas and steam turbines, enhance efficiency and grid stability, delivering continuous power despite intermittent renewable generation.

- **Hydropower:** Hydropower plays a crucial role in stabilizing grids with significant RE. Its ability to rapidly adjust output makes it effective in compensating for wind and solar variability. When renewable generation drops, hydropower meets demand; when abundant, it reduces output to prevent grid overload. This flexibility ensures reliable power supply and grid stability.

11.3.5 Geographical and grid diversification

- **Distributed RES deployment:** Deploying RES across diverse locations reduces regional fluctuations in energy production caused by local weather conditions. For example, reduced solar output in one region due to cloud cover can be offset by higher generation in sunnier areas. This distributed approach enhances reliability, stabilizes RE output, and optimizes resource utilization, making RES a more dependable part of the energy mix.
- **Interconnected power grids:** Interconnecting power grids across regions or countries balances RE variability over large areas. By transmitting surplus energy from regions with high generation to those with higher demand, these networks stabilize supply in real time. They also enable sharing RE across diverse weather patterns and seasons, enhancing reliability and optimizing resource use. This approach supports a more resilient and sustainable energy infrastructure.

11.3.6 Advanced forecasting and predictive tools

- **Weather forecasting:** Advanced weather forecasting technologies are vital for managing RE variability, especially wind and solar power [22]. Accurate predictions of weather conditions help grid operators anticipate changes in wind speeds, solar radiation, and other factors affecting generation. For instance, forecasts indicating reduced solar output due to cloud cover allow operators to adjust energy supply or demand-side strategies. These tools enable proactive planning, enhancing grid stability and efficiency in RE integration.
- **Data analytics and AI:** The use of data analytics and AI is transforming the way RE forecasting and management are approached. ML models scrutinize the historical data, real-time inputs, and environmental factors to improve the accuracy of energy production forecasts. This enables better coordination of supply and demand, enhancing resource efficiency and minimizing imbalances. AI systems also adapt and improve over time, enhancing predictive analysis precision. This advanced forecasting helps grid operators to make decisions in advance, thus reducing reliance on backup power and enhancing grid efficacy and stability.

11.3.7 Summary

Effective management of RES variability is essential for maintaining grid stability. Key solutions include ESS technologies like batteries, hydrogen, and PHS, along with smart grid technologies and DR programs. These approaches help maintain supply and demand balance by accumulating excess energy for periods of low renewable generation, while hybrid systems and flexible power generation methods improve grid resilience and reduce dependence on fossil fuels. These strategies ensure a reliable power supply while promoting sustainable energy practices

11.4 GRID INTERFACING TECHNOLOGIES

The integration of RES into modern power systems depends heavily on effective grid interfacing technologies. These technologies fill the gap between variable and intermittent renewable generation and the stability needs of the conventional grid. A combination of power electronics, system architectures (grid-tied or standalone), and smart metering infrastructures works together to ensure efficient and reliable management of RE in smart grids.

11.4.1 Role of power electronics in grid interfacing

Power electronics form the backbone of grid interfacing for RES. The devices, namely, inverters, converters, and rectifiers, form the backbone of grid interfacing systems, allowing for conversion between AC and DC, frequency and voltage regulation, and synchronization with the grid. For example, solar PV systems generate variable and DC output, which needs to be converted into stable AC power compatible with the grid. The power electronics enable this conversion and manage power quality parameters such as harmonics, reactive power, and voltage fluctuations. In [23], the importance of these devices in shaping modern energy systems was discussed, and it is mentioned that the reliability of the grid depends on the responsiveness and adaptability of power electronic systems.

Beyond basic conversion, modern power electronics are embedded with advanced control systems. These systems help regulate voltage and frequency, provide fault ride-through capabilities, and support grid stability by reacting dynamically to changing loads and generation levels. Reference [24] highlights how control algorithms integrated into inverters allow for functions such as reactive power compensation, synchronization, and voltage support—making power electronics essential for both operational flexibility and maintaining power quality in smart grids.

11.4.2 Grid-tied vs. standalone renewable systems

A key distinction in RE systems is whether they are grid-tied or standalone. Grid-tied systems are directly connected to the utility grid, enabling a bidirectional power flow—users can not only consume electricity from the grid but also inject surplus power back into it. This arrangement encourages net metering and gives opportunities to consumers to act as prosumers and get benefited economically by selling excess energy [25]. Moreover, these systems have better reliability, as they can draw power from the grid during low RE generation periods.

In contrast, standalone systems operate independently of the main grid and are ideal for remote locations where grid accessibility is limited. These systems require robust ESS solutions, namely batteries, to ensure continuous power supply during periods without generation. While standalone systems offer energy autonomy, they are generally more complex and expensive due to their dependence on storage and advanced energy management systems.

11.4.3 Advanced metering infrastructure in grid interfacing

Advanced metering infrastructure (AMI) plays an important role in modern grid interfacing by enabling two-way communication between consumers and utilities. AMI comprises smart meters, communication networks, and data management systems that allow real-time or near real-time tracking of electricity usage. This enables dynamic pricing, DR, and improved load forecasting, all of which are essential for integrating variable RES [26, 27].

Moreover, AMI supports interoperability—the ability of different smart devices and systems to communicate and work together effectively. In the context of smart grids, interoperability ensures that distributed generators, ESS, EVs, and smart appliances can respond to grid signals in a coordinated way. This enhances grid flexibility, reduces peak demand, and supports energy efficiency goals, making AMI a cornerstone of digital energy infrastructure.

Despite the advanced technology, grid interfacing technologies face several challenges including regulatory and standardization gaps, cybersecurity concerns, and the high upfront costs associated with advanced power electronics and communication infrastructure. Additionally, integrating these technologies at scale requires coordinated efforts in policy, market design, and workforce training.

Nonetheless, continuing research and development efforts are addressing these barriers. Innovations in wide bandgap semiconductor materials (like SiC and GaN), edge computing, and machine learning for grid control are poised to make grid interfacing more cost-effective and intelligent. As the global energy transition accelerates, robust and adaptive grid interfacing technologies will be essential for building resilient, efficient, and low-carbon power systems.

11.5 SMART GRID SOLUTIONS FOR INTEGRATION OF RES

The challenges that arise from RES integration into grid can be managed with robust solutions provided by smart grid technologies.

11.5.1 Role of SCADA and IoT in managing renewables

The integration of SCADA and Internet of Things (IoT) technologies plays a pivotal role in the management and optimization of RE systems. Recent studies highlight the deployment of IoT-based architectures to enhance the monitoring, control, and decision-making processes within RE setups. An IoT-based hybrid RE system architecture was developed for a smart campus, comprising wind turbines, PV systems, batteries, and diesel generators [28]. The layered approach, covering power, data acquisition, communication, and application layers, demonstrates how IoT facilitates comprehensive system management. A similar IoT-based SCADA system was designed for monitoring hybrid renewable systems including PV, wind, and battery storage [29]. It illustrated the importance of real-time, cost-effective condition monitoring to optimize asset utilization and prevent blackouts caused by faults. This highlights the important role of IoT-enabled SCADA in providing real-time insights for RE assets.

There is a security concern associated with the SCADA systems. The vulnerabilities of SCADA systems to cyber-attacks were reviewed, and AI-based solutions were proposed to uphold security [30]. This is particularly relevant for RE management, where system integrity is vital for reliable operation. Furthermore, secure SCADA architectures through DNP3 protocol testing were explored, emphasizing the need for robust security frameworks in critical infrastructure [31].

The management of RES benefits from advanced data-driven and machine learning techniques. A decision tree-based prediction approach was developed to improve energy management stability in smart grids, facilitating better forecasting and resource allocation [32]. Similarly, an intelligent control system was presented for district heating, using machine learning and predictive control to optimize energy efficiency and reduce environmental impact [33]. The integration of IoT cloud services further enhances the management of small wind turbines, as demonstrated in [34]. The system leverages IoT cloud platforms for monitoring and controlling turbines, illustrating how cloud-based IoT solutions can support scalable and remote RE asset management.

11.5.2 Microgrids and virtual power plants

The incorporation of RES within microgrids and virtual power plants (VPPs) has garnered significant research interest, emphasizing the role of advanced control strategies and optimization techniques in enhancing grid stability and efficiency [35]. The effectiveness of AI-based energy management systems was demonstrated in microgrids utilizing solar PV technology and energy storage, achieving notable cost savings, i.e., 23.34% on clear days and 45.55% on cloudy days. This highlights the AI's potential in optimizing RES dispatch within grid-connected microgrids. An artificial neural network (ANN)-based binary backtracking search algorithm was utilized for the optimal scheduling of microgrids forming VPPs [36]. It highlights the role of intelligent algorithms in efficiently managing RES integration, ensuring cost-effective operation and sustainability.

A load-frequency control strategy based on a proportional–integral–derivative controller augmented with adaptive fuzzy logic was developed for frequency stability in microgrids with diverse RES, such as wind and PV [37]. This approach enables the microgrid to maintain stability under varying load conditions, demonstrating the critical role of adaptive control in RES-rich environments. A multi-objective particle swarm optimization algorithm was used to optimize shunt active power filters for enhancing power quality [38]. It illustrated how power quality issues, such as harmonic distortion, can be mitigated in PV–wind-based smart grids, thereby supporting RES integration at the distribution level.

The concept of dynamic VPP, introduced in [39], advances the management of RES by incorporating local, global, and economic dynamics. This framework helps in the optimal dispatch of different RES, ensuring flexible and efficient grid integration through real-time adaptation and coordination. A comprehensive review of VPP optimization strategies was provided, focusing on modeling, management, and operational constraints [40]. It highlighted the ongoing research challenges, including the need to balance local resource limitations with more general sustainability objectives. The evolution from traditional microgrids to smart grids and VPP is explored [41] and technological advancements such as smart metering, dynamic pricing, and intelligent energy management systems are discussed. These innovations support two-way communication and decentralized control, address key operational challenges, and enhance the resilience of RES incorporation. In the context of market participation, a p-robust method to minimize the negative impacts of wind generation deviations on electricity markets was presented [42]. This approach improved the reliability of wind integration within VPPs and added to stable and efficient energy markets.

Reinforcement learning (RL) is also gaining momentum in this domain. An RL-based control strategy for EV and RES-driven VPPs is investigated, which demonstrated how machine learning can enable continuous, adaptive optimization under dynamic grid conditions [43]. In [44], the integration of RES in microgrids and smart grids was compared within deregulated power systems, highlighting the importance of flexible, intelligent grid in meeting energy demand for both urban and rural areas.

11.5.3 DR and ESS

Incorporating RES into smart grids necessitates effective DR strategies and ESS to ensure stability, reliability, and optimal operation. Several studies highlighted the important role of these components in managing the variability and intermittency associated with RES.

A comprehensive overview of DR strategies within smart home energy management systems was presented [45] and highlighted their importance in balancing energy consumption with RE generation. It also discussed various DR approaches and the integration of EVs and RES, demonstrating how demand flexibility can enhance the RES utilization. An AI-based energy management system integrating energy storage to optimize microgrid dispatch was introduced [35], focusing on solar PV system. This approach demonstrated significant cost savings, and indicated that the intelligent control of ESS can manage RES integration by reducing supply fluctuations and operational costs. The optimal sizing of PV-battery hybrid systems in smart homes was analyzed considering grid outages and DR [46]. This methodology provides guidelines to implement ESS for improved reliability and resilience, especially during grid disturbances, while ensuring continuous RES utilization. The use of model predictive control combined with multi-objective optimization was explored to uphold frequency stability in smart grids with integrated PV, wind, fuel cells, and batteries [47]. This hybrid control approach accounted for the challenges posed by unpredictable RE generation, ensuring stable grid operation through coordinated energy storage management.

Operational planning in smart grids and advanced control techniques for distributed energy storage were reviewed [48]. It classified control strategies namely automatic voltage regulation and load frequency control, highlighting their significance in maintaining grid stability amid high-RES penetration. A smart home energy management architecture was developed that incorporates ESS and DR programs [49]. The model presented emphasized secure communication and knowledge integration, resulting in effective RES management at the residential level and enhanced system flexibility. In the context of market operations, a robust method for integrating DR and wind power aggregators with generation units and storage in wholesale markets was developed [42]. This approach minimized the negative impacts of wind variability, illustrating how coordinated DR and ESS can stabilize market participation and RES integration. Microgrid operation under critical conditions was examined [50], and it highlighted the role of DR and ESS in optimizing performance by minimizing costs and load shedding. This finding reinforced the importance of ESS in maintaining microgrid resilience during operational challenges.

In [51], focus was on bidirectional charging systems for vehicle-to-grid (V2G) and grid-to-vehicle (G2V) interactions. It proposed methods to enhance energy transfer efficiency and storage management. This work illustrated the potential of EVs in mobile energy storage that supports RES integration and grid stability. Finally, the challenges of integrating RES into distribution networks were addressed [52]. A two-level optimization model aimed at improving network resilience and operational efficiency was presented, which emphasized the necessity of advanced control and storage solutions to handle RES variability.

11.6 CASE STUDIES

The integration of RE into smart grids differs significantly across countries, shaped by differences in their governance models, regulatory structures, market liberalization, technological innovation, and institutional capacities. This section presents a series of real-world case studies that illustrate diverse strategies for RE integration within smart grids. These cases highlight the technological advancements and reveal the governance challenges, policy innovations, and stakeholder collaborations that influence the success of smart grid transitions.

11.6.1 Case 1: China and Japan—a comparative study

A comparative case study of China and Japan provides unique insights into how state-market dynamics influence the deployment of smart grid technologies for RE integration [53]. Both countries have demonstrated strong commitments to climate goals and energy transition, yet their institutional arrangements, regulatory environments, and implementation strategies have diverged significantly.

11.6.1.1 China: state-led grid expansion and limited market liberalization

China has embraced a top-down, incumbent-led smart grid strategy that has been largely orchestrated by the two state grids: the State Grid Corporation of China and China Southern Power Grid. The nation has made substantial investments in its super-grids and ultra-high voltage transmission infrastructure to support large-scale renewables such as wind and solar. However, limited retail market opening and support for new private sector entrants have inhibited innovation in demand-side response and decentralized energy. The main features of China's model are:

- Centralization of investment through the state-owned enterprises.
- Strong emphasis on hardware infrastructure (smart meters, transmission lines, etc.).
- Regulatory obstacles to non-governmental entities, such as delay in grid access and lack of clear price incentives.
- Lack of policy provision to residential DR or prosumer.
- Ongoing challenges associated with the curtailment of RE resources (wind and solar), owing to grid congestion and the absence of ancillary service mechanism.

Despite these challenges, some public–private cooperation is evolving and state-owned banks are supporting solar projects with loans and insurance.

11.6.1.2 Japan: community-oriented smart grid demonstrations and market reforms

In contrast, Japan has followed a bottom-up, community-based model, largely shaped by post-Fukushima energy concerns. The country's smart grid progress has been characterized by four major smart community demonstration projects in cities like Yokohama, Toyota, and Kitakyushu, led by local governments, research institutions, and private technology firms.

Japan's approach emphasizes:

- Integration of DER and energy storage.
- A more liberalized electricity market in retail, which was fully opened in 2016.
- Use of smart meters and dynamic pricing to encourage consumer engagement and demand-side management (DSM).
- A collaborative governance model that involves multiple stakeholders with local governments, technology providers, and academia.
- Experimentation with policy learning and real-time feedback from demonstration sites.

While Japan still has vertically integrated utilities in the transmission and distribution sectors, reforms have encouraged more market players and innovative business models. However, challenges are there in scaling up the model into national grid planning.

11.6.1.3 Comparative insights

The study concludes that both China and Japan acknowledge the importance of government-market coordination, but apply it differently:

- China relies heavily on state-owned monopolies, resulting in rapid infrastructure deployment but constrained innovation.
- Japan adopts decentralized experimentation and market diversity, leading to slower deployment but better innovation ecosystems.

Five key functions shape these dynamics:

- Market formation: driven by state mandates in China; by competitive incentives and local initiatives in Japan.
- Market regulation: centralized in China; liberalized and increasingly participatory in Japan.
- Management of public goods: both countries invest in R&D and standardization, but Japan is more transparent in data sharing and stakeholder involvement.
- Networking and resource mobilization: strong state-industry ties in China vs. broader public–private–academic collaboration in Japan.
- Policy learning: limited in China but actively pursued in Japan through structured feedback loops.

11.6.1.4 Conclusion

This comparative case illustrates how national context—especially political structure, market liberalization, and institutional culture—deeply affects the trajectory of smart grid development and RE integration. The developing countries boarding on similar transitions can take lessons from China and Japan on the trade-offs between centralized efficiency and decentralized innovation, and the importance of aligning governance frameworks with socio-technical goals.

11.6.2 Case 2: Oman—transitioning to smart grids through renewable integration

Oman's journey toward RE integration within its grid infrastructure reflects the country's broader goals of diversifying energy resources, enhancing energy security, and achieving environmental sustainability. With ample solar irradiance and growing electricity demand, Oman has recognized the need to modernize its power system through smart grid technologies and optimized RE deployment.

This study evaluates multiple renewable integration scenarios using techno-economic models to assess the impact of solar PV and wind power on Oman's grid [54]. The analysis includes both centralized and distributed configurations, investigating their implications on operational costs, atmospheric emissions, losses, and system efficiency. The main findings of this case study are:

- Solar PV emerges as the most viable renewable option owing to the country's high solar potential and the decreasing cost of PV systems.
- Integrating RE into the existing grid infrastructure can significantly reduce fuel consumption and greenhouse gas emissions.
- Smart grid features—namely AMI, distributed energy management systems, and real-time monitoring—are indispensable for managing the variability of RES.
- Significant economic benefits, including reduced levelized cost of electricity (LCOE) and operational cost savings after incorporating smart grid technology.

11.6.2.1 Policy and planning considerations

Oman's electricity market is undergoing restructuring under the supervision of the Authority for Public Services Regulation. The development of an open and competitive electricity market, along with regulatory support for independent power producers, is crucial for accelerating renewable deployment. The study emphasizes:

- The need for strong regulatory frameworks to support net metering, rooftop solar adoption, and grid code updates.
- Strategic investment in smart infrastructure for effective DSM and integration of distributed generation.
- A phased approach for smart grid implementation, initially implementing the pilot projects and then scaling up across regions.

11.6.2.2 Challenges and future outlook

While Oman has made progress in initiating RE projects—such as the Ibri solar plant and wind farms in Dhofar—certain challenges remain:

- Grid flexibility limitations to handle variable output from RES
- High initial capital investment in smart technologies
- Need for capacity building and technical training to operate digital infrastructure
- Consumer engagement and awareness for DR initiatives

However, the country's planned direction and ongoing reforms are creating a favorable environment for smart grid evolution. The study acclaims adoption of integrated planning approaches, combining RE expansion with digital grid modernization to ensure a stable, cost-efficient, and sound power system.

11.6.2.3 Summary

Oman's case reflects the potential of smart grids for a clean energy transition in resource-rich, fossil-dependent economies. By leveraging its solar resources and smart technologies, Oman is laying the foundation for a resilient and future-ready power sector. Its approach, although still in early stages, serves as a model for developing countries to diversify their energy mix through sustainable innovation.

11.6.3 Case 3: Philippines—RE integration and grid interfacing in off-grid islands

The Philippines, with more than 7,000 islands, faces a unique challenge in providing reliable, affordable, and clean electricity to its numerous off-grid communities. Numerous of these islands rely on costly diesel-based generation, making them prime candidates for RE integration supported by smart grid technologies.

A recent techno-economic assessment focused on 502 sample off-grid islands, applying cluster analysis to group them based on socio-economic, physical, and RE potential parameters [55]. The aim was to allow fast-track design of tailored smart energy systems and facilitate the imitation of proven solutions across similar islands. Islands were first categorized into five population-based groups, with additional in-depth clustering applied to small and medium-sized islands. The analysis revealed that similarities between islands are not necessarily linked to geographic proximity, highlighting the importance of a national-scale classification approach for energy system planning. The results showed:

- Island size and demand profiles: Most islands fall into very small or small categories, often lacking distinctive load profiles or opportunities for sophisticated DSM. In such cases, simple hybrid systems—combining solar PV, battery storage, and diesel backup—were identified as the most practical solutions. For non-remote islands with higher population density, grid interconnection with nearby main islands can be more economical and technically viable.
- Renewable potential: Solar energy emerged as the most suitable renewable source for almost all islands, even where land

constraints exist. Wind potential was more site-specific, with varying feasibility depending on local climatic conditions. The study suggests solar PV–battery systems as the backbone of most off-grid island power solutions, with wind integration possible in select locations.

- Economic diversity and DSM opportunities: Medium to large islands showed diverse economic activities, including trade, commerce, and industry, which present higher DSM potential. Tourist-centric islands were identified as prime candidates for cogeneration systems and thermal energy management due to their substantial cooling and heating demands.
- Energy–water nexus potential: Islands with limited freshwater resources could benefit from integrating desalination and water supply systems into their smart energy plans, creating synergies between energy production and water security.

11.6.3.1 Grid interfacing and system design implications

The study emphasizes that renewable integration strategies must be matched with suitable grid interfacing technologies:

- Hybrid PV–battery–diesel microgrids for small, isolated islands.
- Advanced inverter systems to maintain power quality and stability.
- Smart controllers for real-time balancing of supply and demand.
- Load control systems for DSM in larger islands with commercial and industrial demand.

11.6.3.2 Policy and implementation recommendations

To accelerate deployment, the study recommends:

- Regulatory reforms to address permitting and interconnection challenges in remote areas.
- Cluster-based project development, enabling private sector participation by offering groups of similar islands under unified contracts.
- Financial support mechanisms to reduce the high upfront costs of renewable and storage technologies.
- Inclusion of energy–water nexus solutions in long-term island electrification planning.

11.6.3.3 Conclusion

This case study highlights that the Philippines' pathway to sustainable off-grid electrification lies in context-specific renewable integration paired with flexible grid interfacing solutions. By leveraging solar PV, targeted wind projects, and smart microgrid designs, and by clustering islands for streamlined project execution, the country can transition rapidly toward smarter, cleaner, and more resilient island energy systems.

11.6.4 Case 4: India (Jodhpur)—grid-connected solar integration and performance analysis

India's RE ambitions are strongly supported by regions with high solar potential, such as Jodhpur in Rajasthan, which receives some of the highest annual solar irradiation levels in the country. This case study focuses on the performance assessment of a grid-connected solar PV system in Jodhpur, highlighting the technical, economic, and operational aspects of integrating large-scale solar into the regional grid.

The system under study consists of a grid-connected solar PV installation designed to export electricity directly to the local distribution network [56]. The objectives were to:

- Evaluate energy production and performance ratio under Jodhpur's climatic conditions.
- Assess seasonal variations in generation.
- Analyze the economic viability of grid-connected solar in high-irradiance desert environments.

The following are the main findings of this case study:

1. Energy generation and performance ratio
   - The PV plant achieved a high annual capacity factor, attributed to consistent solar radiation and low cloud cover.
   - Performance ratio values were within the expected range for desert climates, although minor losses were noted due to soiling from dust storms, which is a common issue in the Thar Desert.
2. Seasonal and diurnal patterns
   - Peak generation occurred in winter months due to cooler ambient temperatures, which improve PV module efficiency, despite slightly lower solar intensity compared to summer.
   - Summer months showed marginal efficiency reductions due to high module temperatures, even though irradiation was at its peak.
3. Grid interfacing and stability considerations
   - The system used string inverters with maximum power point tracking (MPPT) to optimize output under varying sunlight conditions.

- Grid compliance was ensured through synchronization and anti-islanding protection features.
- The study emphasized the need for reactive power control and voltage support in areas with high PV penetration to maintain grid stability.

4. Economic analysis
   - The LCOE was found to be competitive with conventional sources, supported by government incentives and falling PV module prices.
   - The project demonstrated strong return on investment potential in similar high-solar regions.

11.6.4.1 Challenges and mitigation

- Soiling losses: Regular cleaning schedules were recommended to reduce output losses from dust accumulation.
- Temperature-induced efficiency drop: The use of bifacial modules or improved ventilation design was suggested to mitigate thermal losses.
- Grid voltage fluctuations: Smart inverters with voltage regulation capability were highlighted as a necessary upgrade for future expansion.

11.6.4.2 Implications for RE integration

The Jodhpur case confirms that grid-connected PV systems in high-irradiance zones can deliver reliable and cost-effective power when paired with robust grid interfacing technologies. Lessons from this project include:

- Incorporating advanced inverter controls for grid stability.
- Aligning maintenance schedules with local environmental conditions.
- Scaling deployment in other arid and semi-arid regions with similar solar profiles.

11.6.4.3 Conclusion

Jodhpur's solar PV integration illustrates the technical feasibility and economic viability of large-scale solar projects in India's desert states. The combination of favorable natural resources, supportive policy measures, and mature grid interfacing technologies positions such regions as strategic hubs for India's RE expansion. The findings reinforce the role of site-specific performance analysis in guiding future solar deployment strategies (Table 11.1).

Table 11.1 Summary of case studies on RE integration and grid interfacing

Case study	*Key renewable sources*	*Integration approach*	*Notable outcomes*	*Main challenges*	*Key lessons*
China–Japan [55]	Solar PV, Wind	China: Centralized, state-driven grid modernization; Japan: Decentralized, community-based smart grids with stakeholder participation	Large-scale renewable deployment in China; successful localized demonstrations in Japan	China: Managing grid stability with high-RES share; Japan: Scaling up localized pilots	Integration strategies must match governance structures; both centralization and decentralization can succeed if aligned with context
Oman [56]	Solar PV	Hybrid renewable systems with smart grid features, centralized and distributed configurations	Reduced fuel dependency, lower emissions, improved LCOE	High capital costs, grid flexibility limits, capacity building needs	Phased smart grid upgrades and clear regulations are essential for RES transition in fossil-reliant economies
Philippines [57]	Solar PV, Wind, Small Hydro	Cluster-based microgrid design, PV–battery–diesel hybrids, DERMS, AMI	Tailored solutions for 502 off-grid islands; potential energy–water nexus integration	Limited funding, technical capacity in rural utilities, complex multi-island grid management	Clustering similar islands accelerates deployment; integrating water systems can add socio-economic value
Jodhpur, India [58]	Solar PV	Grid-connected PV plant with advanced inverter controls, MPPT, anti-islanding	High performance ratio, competitive LCOE, strong ROI potential	Soiling losses, temperature-induced efficiency drop, voltage fluctuations	Maintenance schedules and smart inverter functions are critical for desert solar projects

11.6.5 Summary

The study of RE integration and grid interfacing in China–Japan, Oman, the Philippines, and Jodhpur (India) highlights how local conditions, governance styles, and market structures influence these technology choices as well as implementation strategies. China's state-led, large-scale grid modernization contrasts with Japan's localized, stakeholder-driven smart grid trials, each offering distinct strengths in scaling renewables. Oman demonstrates the value of strategic planning, regulatory reform, and gradual smart grid adoption to enable a transition toward solar-based hybrid systems. In the Philippines, a cluster-based evaluation of off-grid islands highlights the need for location-specific microgrids and PV–battery configurations, with added opportunities in linking energy systems to water supply solutions. The Jodhpur case showcases the strong performance of grid-connected solar in high-irradiance environments, emphasizing inverter technology, voltage regulation, and maintenance for optimal operation. Together, these cases demonstrate that successful integration of renewables requires tailored system design, high-quality grid interconnections, suitable policies, and active involvement of all stakeholders.

11.7 CONCLUSION

The incorporation of RES into modern power systems is both a necessity and a challenge in the global transition toward sustainable, low-carbon energy. This chapter examined that while RES namely solar, wind, hydro, biomass, geothermal, ocean energy, and fuel cells offer substantial environmental and economic benefits, their variability demands robust strategies for ensuring reliability, stability, and efficacy. Smart grids, supported by advanced control systems, power electronics, energy storage, and intelligent DSM, emerge as the key enabler for effective RE integration. Grid interfacing technologies, including smart inverters, AMI, and real-time communication platforms, bridge the gap between intermittent RE generation and the operational requirements of conventional grids. Complementary solutions such as hybrid energy systems, forecasting tools, and microgrids further enhance adaptability and resilience. Case studies from China, Japan, Oman, the Philippines, and India highlight that successful RE integration depends not only on technological readiness but also on governance structures, regulatory frameworks, market mechanisms, and stakeholder engagement.

The overall summary of RE integration in smart grid is presented in Figure 11.3. Ultimately, the path forward requires context-specific approaches that combine technical innovation with supportive policies and active community participation. By aligning renewable deployment with smart grid advancements and tailored integration strategies, power systems

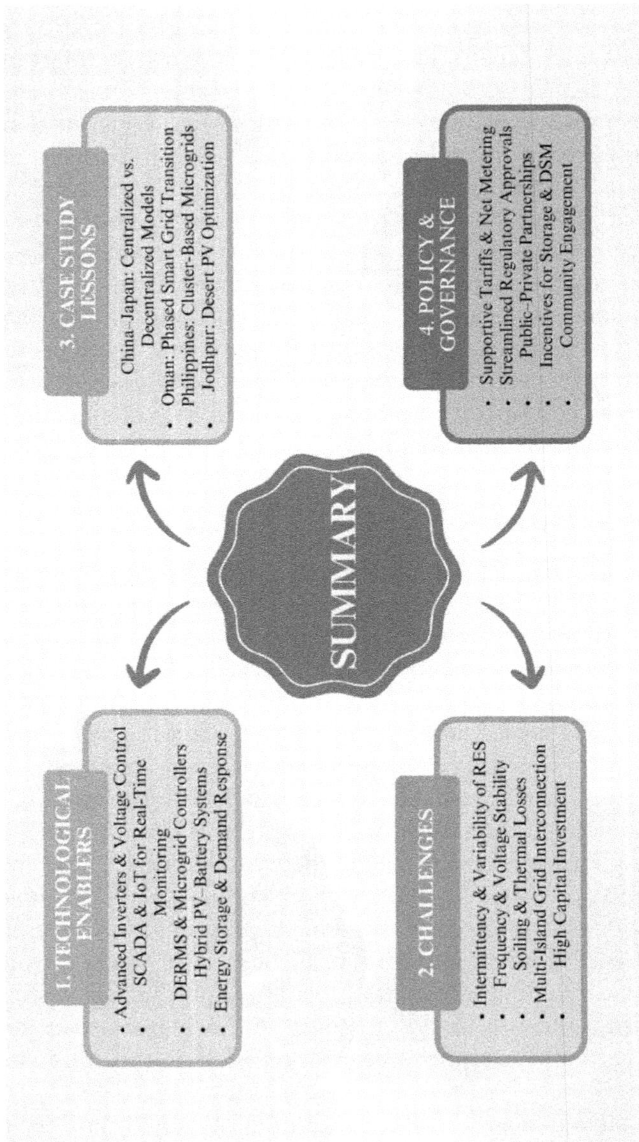

Figure. 11.3 Summary of renewable energy integration and grid interfacing.

can achieve higher RE penetration while maintaining stability, improving energy access, and advancing global sustainability goals.

REFERENCES

[1] S. Sagar, U. Yadav, N. Bansal, and J. Kumar, "Analyzing the Effects of Renewable Energy Resources on Smart Grid Development," International Conference on Cutting-Edge Developments in Engineering Technology and Science, pp. 275–280, 2024. doi: 10.62919/ncxs1248.

[2] M. Anan, Integrating Renewable Energy Sources into Smart Grids: Opportunities and Challenges, in *Renewable Energy Technologies for Water Desalination*, 1st ed., CRC Press, 2017, p. 10. ISBN: 9781315643915.

[3] F. Ayadi, I. Colak, I. Garip, and H. Bulbul, "Impacts of Renewable Energy Resources in Smart Grid," in *Proc. Int. Conf. Smart Grid*, Jun. 2020, pp. 183–188. doi: 10.1109/icSmartGrid49881.2020.9144695.

[4] G. Nalcaci, N. Yarar, and G.-W. Weber, "Potential Sources of Renewable Energy in Smart Grids: Opportunities and Challenges," In: Advancements in Renewable Energy and Green Hydrogen, pp. 1–26, Feb. 2024. doi: 10.4018/979-8-3693-1014-4.ch001.

[5] G. Anusha, K. Arora, H. Sharma, S. P Thota, G. P. Joshi, and W. Cho, "Control Strategies of 15-Level Modified Cascaded H-Bridge MLI with Solar PV and Energy Storage System," *Energy Rep.*, 12, 2–26, 2024. https://doi.org/10.1016/j.egyr.2024.06.003.

[6] T. V. Krishna, M. K. Maharana, and C. K. Panigrahi, "Challenges in Integrating Renewable Sources into a Utility Grid." *International Journal of Innovative Technology and Exploring Engineering*, vol. 9, pp. 91–97, 2020. https://doi.org/10.35940/ijitee.c8194.019320.

[7] H. Nanda, H. Sharma, K. Arora, A. Yadav, G. P. Joshi, and W. Cho, "A Novel High-Gain Switched-Capacitor Multilevel Inverter with Reduced Components for Grid Integration," *Alex. Eng. J.*, vol. 106, pp. 240–251, 2024. https://doi.org/10.1016/j.aej.2024.03.067.

[8] H. Javed, P. H. Nguyen, J. Morren, and J. H. Slootweg, "Review of Operational Challenges and Solutions for DER Integration with Distribution Networks," in *Proc. Int. Univ. Power Eng. Conf. (UPEC)*, 2021, pp. 1–6. doi: 10.1109/UPEC50034.2021.9548198.

[9] D. P. Kothari, R. Ranjan, and K. Singal, *Renewable Energy Sources and Emerging Technologies*. PHI Learning, 2021.

[10] M. K. G. Deshmukh et al., "Renewable Energy in the 21st Century: A Review," *Materials Today: Proc.*, vol. 80, pp. 1756–1759, 2023. doi: 10.1016/j.matpr.2021.05.501.

[11] M. Farghali, A. I. Osman, and Z. Chen et al., "Social, Environmental, and Economic Consequences of Integrating Renewable Energies in the Electricity Sector: A Review," *Environ. Chem. Lett.*, vol. 21, pp. 1381–1418, 2023. doi: 10.1007/s10311-023-01587-1.

[12] A. Aldosary, "The Role of Sodium-Nickel Chloride (Na-$NiCl_2$) Batteries in Managing Uncertainty and Renewable Sources for Empowering Hybrid Energy Systems Using Bi-Level CONOPT-Based Optimization,"

Ain Shams Eng. J., vol. 16, no. 1, pp. 103174, 2025. doi: 10.1016/j.asej.2024.103174.

[13] R. N. Vitorino et al., "Pumped-Hydro Storage Plants Influencing the Optimal Hourly Dispatch in Hydro-Thermal Systems with Massive Integration of Variable Renewable Energy," *Electr. Power Syst. Res.*, vol. 220, p. 109282, 2023. doi: 10.1016/j.epsr.2023.109282.

[14] S. W. Monie and M. Åberg, "Potential to Balance Load Variability, Induced by Renewable Power, Using Rock Cavern Thermal Energy Storage, Heat Pumps, and Combined Heat and Power in Sweden," *Appl. Energy*, vol. 343, pp. 121210, 2023. doi: 10.1016/j.apenergy.2023.121210.

[15] G. Yan et al., "Solar-Powered Compact Thermal Energy Storage System with Rapid Response Time and Rib-Enhanced Plate via Techniques of CFD, ANN, and GA," *J. Energy Storage*, vol. 105, pp. 114807, 2025. doi: 10.1016/j.est.2024.114807.

[16] R. Chen, D. Lu, T. Shen, and M. Gong, "A New Dynamic Control Strategy for a Solar-Driven Absorption Thermal Energy Storage System: Modeling and Performance Evaluation," *Appl. Therm. Eng.*, vol. 260, pp. 124983, 2025. doi: 10.1016/j.applthermaleng.2024.124983.

[17] H. Sharma, J. Bharali, M. Motghare et al. "Development and Evaluation of Hybrid Harris Hawks Optimization Algorithms for Advanced Engineering Applications," *Sci Rep*, vol. 15, p. 43123, 2025. https://doi.org/10.1038/s41598-025-23624-5.

[18] M. Akbarizadeh, T. Niknam, and H. Jokar, "Economic Operation of Networked Flexi-Renewable Energy Hubs with Thermal and Hydrogen Storage Systems Based on the Market Clearing Price Model," *Int. J. Hydrogen Energy*, vol. 50, pp. 1–18, 2024. doi: 10.1016/j.ijhydene.2023.06.144.

[19] B. Wang and B. Zhaoxiang, "Hydrogen Energy Storage: Mitigating Variability in Wind and Solar Power for Grid Stability in Australia," *Int. J. Hydrogen Energy*, vol. 97, pp. 1249–1260, 2025. doi: 10.1016/j.ijhydene.2024.11.491.

[20] Y. W. Koholé, C. A. W. Ngouleu, F. C. V. Fohagui, and G. Tchuen, "A Comprehensive Comparison of Battery, Hydrogen, Pumped-Hydro and Thermal Energy Storage Technologies for Hybrid Renewable Energy Systems Integration," *J. Energy Storage*, vol. 93, p. 112299, 2024. doi: 10.1016/j.est.2024.112299.

[21] A. Sharma and S. K. Jain, "Procurement of Electrical Energy and Operating Reserve Ancillary Services in Wind Incorporated Deregulated System Considering Demand Response," *IET Gener. Transm. Distrib.*, vol. 14, no. 15, pp. 3006–3021, 2020. doi: 10.1049/iet-gtd.2019.0679.

[22] H. Wang et al., "Taxonomy Research of Artificial Intelligence for Deterministic Solar Power Forecasting," *Energy Convers. Manage.*, vol. 214, p. 112909, 2020. doi: 10.1016/j.enconman.2020.112909.

[23] F. Blaabjerg, Y. Yang, K. Ma, and X. Wang, "Advanced Grid Integration of Renewables Enabled by Power Electronics Technology," in *Renewable Energy Devices and Systems with Simulations in MATLAB and ANSYS*, 2015, pp. 3–9. doi: 10.1007/978-3-658-10958-5_1.

[24] V. Monteiro, T. Soares, J. P. Lopes, M. A. Matos, and J. L. Afonso, "Interfacing Power Electronics Systems for Smart Grids: Innovative Perspectives of Unified Systems and Operation Modes," in *Proc. IEEE Conf. Ind. Electron. Soc. (IECON)*, 2021, pp. 1–6. doi: 10.1109/IECON48115.2021.9589903.

[25] R. Mishra, U. Yadav, H. Sah, and J. Kumar, "Synergizing Smart Grid Technology and Renewable Energy Systems," in *Proc. Conf.*, May 2024, pp. 209–215. doi: 10.62919/xzxz2139.

[26] I. Poursanidis, E. Kotsakis, N. Andreadou, and M. Masera, "Evaluation of Interoperability in the Context of Advanced Metering Infrastructure," in *Proc. IEEE Int. Conf. Smart Energy Grid Eng. (SEGE)*, 2018, pp. 121–127. doi: 10.1109/SEGE.2018.8499375.

[27] J. F. Martins, A. G. Pronto, V. Delgado-Gomes, and M. Sanduleac, "Smart Meters and Advanced Metering Infrastructure," in *Pathways to a Smarter Power System*, A. Taşcıkaraoğlu and O. Erdinç, Eds. Academic Press, 2019, pp. 89–114. doi: 10.1016/B978-0-08-102592-5.00004-1.

[28] A. M. Eltamaly, M. A. Alotaibi, A. I. Alolah, and M. A. Ahmed, "IoT-Based Hybrid Renewable Energy System for Smart Campus," *Sustainability*, vol. 13, p. 8555, 2021. https://doi.org/10.3390/su13158555.

[29] L. A. Aldossary, M. Ali, and A. Alasaadi, "Securing SCADA Systems against Cyber-Attacks Using Artificial Intelligence," in *2021 International Conference on Innovation and Intelligence for Informatics, Computing, and Technologies (3ICT)*, 2021.

[30] M. O. Qays et al., "Monitoring of Renewable Energy Systems By IoT-aided SCADA System," *Energy Sci. Eng.*, vol. 10, pp. 1874–1885, 2022. https://doi.org/10.1002/ese3.1130.

[31] A. Soliman et al., "Towards A Secured SCADA Architecture: A DNP3 Test Case," in *Proc. IEEE SoutheastCon*, 2023.

[32] S. Chen et al., "Decision Tree-based Prediction Approach for Improving Stable Energy Management in Smart Grids," *J. High Speed Netw.*, vol. 29, 2023.

[33] D. Stojiljković et al., "Data-driven Approaches for Intelligent Control in District Heating Systems," in *Proc. SAUM*, 2024.

[34] M. Abuzaid, J. Iqbal, and V. Singhal, "IoT-Based Monitoring, Communication, and Control of Small Wind Turbines Using IoT Cloud Service," *Int. Res. J. Comput. Sci.*, vol. 11, pp. 159–164, 2024.

[35] A. Kumar, M. Alaraj, M. Rizwan, and U. Nangia, "Novel AI Based Energy Management System for Smart Grid with RES Integration," *IEEE Access*, vol. 9, pp. 162530–162542, 2021. https://doi.org/10.1109/ACCESS.2021.3131502.

[36] M. A. Hannan et al., "ANN-Based Binary Backtracking Search Algorithm for VPP Optimal Scheduling and Cost-Effective Evaluation," *IEEE Trans. Ind. Appl.*, 2021.

[37] A. A. A. El-Gammal and C. Boodoo, "Optimal Frequency Stability Control Strategy for A Grid-Connected Wind/PV/FC/BESS Coordinated with Hydroelectric Power Plant Storage Energy System Using Variable Structure Control," *Eur. J. Energy Res.*, 2021.

[38] A. Elgammal and T. Ramlal, "Minimum Harmonic Distortion Losses and Power Quality Improvement of Grid Integration Photovoltaic-Wind Based Smart Grid Utilizing MOPSO," *Eur. J. Energy Res.*, vol. 1, pp. 23–29, 2021. https://doi.org/10.24018/ejenergy.2021.1.4.26.

[39] B. Marinescu, O. Gomis-Bellmunt, F. Dörfler, H. Schulte, and L. Sigrist, "Dynamic Virtual Power Plant: A New Concept for Grid Integration of Renewable Energy Sources," *IEEE Access*, vol. 10, pp. 104980–104995, 2021. https://doi.org/10.1109/ACCESS.2022.3205731.

[40] B. Goia, T. Cioara, and I. Anghel, "Virtual Power Plant Optimization in Smart Grids: A Narrative Review," *Future Internet*, vol. 14, p. 128, 2022. https://doi.org/10.3390/fi14050128.

[41] F. Badal et al., "Microgrid to Smart Grid's Evolution: Technical Challenges, Current Solutions, and Future Scopes," *Energy Sci. Eng.*, vol. 11, pp. 874–928, 2022. https://doi.org/10.1002/ese3.1319.

[42] R. Nourollahi and R. Esmaeilzadeh, "Integration Demand Response Aggregator and Wind Power Aggregators with Generation Units and Energy Storages in Wholesale Electricity Markets," *arXiv*, 2023.

[43] F. Maldonato and I. Hadachi, "Reinforcement Learning Control Strategies for Electric Vehicles and Renewable Energy Sources Virtual Power Plants," *arXiv*, 2024.

[44] S. Dawn et al., "Integration of Renewable Energy in Microgrids and Smart Grids in Deregulated Power Systems: A Comparative Exploration," *Adv. Energy Sustain. Res.*, vol. 5, p. 2400088, 2024. https://doi.org/10.1002/aesr.202400088.

[45] A. Badar and A. Anvari-Moghaddam, "Smart Home Energy Management System – A Review," *Adv. Build. Energy Res.*, vol. 16, 118–143, 2020. https://doi.org/10.1080/17512549.2020.1806925.

[46] M. Tostado-Véliz, D. Icaza-Alvarez, and F. Jurado, "A Novel Methodology for Optimal Sizing Photovoltaic-battery Systems in Smart Homes Considering Grid Outages and Demand Response," *Renew. Energy*, vol. 170, pp. 884–896, 2021. https://doi.org/10.1016/j.renene.2021.02.006.

[47] A. Elgammal and T. Ramlal, "Optimal Model Predictive Frequency Control Management of Grid Integration PV/Wind/FC/Storage Battery Based Smart Grid Using Multi Objective Particle Swarm Optimization MOPSO," *WSEAS Trans. Electron.*, vol. 12, pp. 46–54, 2021. https://doi.org/10.37394/232017.2021.12.7.

[48] M. Jayachandran, C. R. Reddy, S. Padmanaban, and A. H. Milyani, "Operational Planning Steps in Smart Electric Power Delivery System," *Sci. Rep.*, vol. 11, article 17250, 2021. https://doi.org/10.1038/s41598-021-96769-8.

[49] P. Gonzalez-Gil, J. A. Martínez, and A. Skarmeta, "A Prosumer-Oriented, Interoperable, Modular and Secure Smart Home Energy Management System Architecture," *Smart Cities*, vol. 5, pp. 1054–1078, 2022. https://doi.org/10.3390/smartcities5030053.

[50] A. M. Ali, A. Chammam, X. Wang, X. Zhang, F. Zhou, and X. Xu, "Modeling Smart Electrical Microgrid with Demand Response and Storage Systems for Optimal Operation in Critical Conditions," *Sci. Technol. Energy Transit.*, 2024.

[51] M. I. Hossain et al., "Enhancing Performance of Bidirectional Charging System for V2G and G2V in Smart Grid," in *Proc. IEEE Int. Conf. Power, Electrical*, 2024.

[52] P. Singh, K. Arora, and U. C. Rathore, Control Strategies for Improvement of Power Quality in Grid Connected Variable Speed WECS with DFIG–An Overview. *J. Phy.: Conf. Ser.*, vol. 2327, no. 1, p. 012008, 2022. https://doi.org/10.1088/1742-6596/2327/1/012008.

[53] D. N.-Y. Mah, "Conceptualising Government-Market Dynamics in Socio-Technical Energy Transitions: A Comparative Case Study of Smart Grid Developments in China and Japan," *Geoforum*, vol. 108, pp. 148–168, 2020. doi: 10.1016/j.geoforum.2019.07.025.

[54] A. S. Malik et al., "Smart Grid Scenarios and Their Impact on Strategic Plan—A Case Study of Omani Power Sector," *Sustain. Cities Soc.*, vol. 37, pp. 213–221, 2018. doi: 10.1016/j.scs.2017.11.015.

[55] H. Meschede et al., "On the Transferability of Smart Energy Systems on Off-Grid Islands Using Cluster Analysis – A Case Study for the Philippine Archipelago," *Appl. Energy*, vol. 251, p. 113290, 2019. doi: 10.1016/j.apenergy.2019.05.093.

[56] V. K. Saini, R. Kumar, A. S. Al-Sumaiti, and E. Heydarian-Forushani, "Learning Based Short Term Wind Speed Forecasting Models for Smart Grid Applications: An Extensive Review and Case Study," *Electr. Power Syst. Res.*, vol. 222, p. 109502, 2023. doi: 10.1016/j.epsr.2023.109502.

Chapter 12

Design and development of an electric vehicle go-kart

Sachin Chawla, Kirti, Abhishek Bajaj, Shippu Sachdeva, and Keshav Kumar

12.1 INTRODUCTION

12.1.1 Context and motivation

A go-kart is classified as an economical small-scale racing vehicle, thus having a unique set of requirements (structural, driving, and economical). These requirements entail being as economically feasible as possible, and to include elements of strength, rigidity, use of standard parts, and optimization of drive systems, as well as being efficient. Conventional go-kart configurations have relied on small two-stroke internal combustion engines (ICEs) in the 50–125 cc range. While this option is admittedly the cheapest, it must also be noted that it comes with significant downsides regarding environmental sustainability, noise, and inefficiency. Going electric is not only an option, but the most progressive alternative. Most go-karting facilities have been impacted by climate concerns, environmental sustainability, and the advancements in electric motors and batteries. The rapid development and unrelenting emissions of economically powered vehicles also provided the research and regulatory imperative for the development of an environmentally sustainable option for powering go-karts. This most importantly requires the development of a tandem of a lightweight, robust chassis, as well as extensive analysis of various types (torsional, frontal, side, rear impact, stress). In contrast with ICE-powered go-karts, electrics have the advantage of zero emissions, reduced maintenance overhead, noise, and while still being able to provide the same torque at any speed, can also charge from renewable sources.

12.1.2 Design philosophy and objectives

Synthesis of multiple engineering disciplines, including mechanical engineering, electrical engineering, control systems engineering, and materials science, is required for the construction of an electric go-kart. Design objectives include four fundamental principles permeating every part of

 DOI: 10.1201/9781003689430-12

the vehicle: durability, safety and ergonomic features, lightweight construction, and high-performance levels, with an integrated consideration of all parameters, including reliability, safety, cost, performance, aesthetic, and ergonomic pursuits.

The major design goals for the construction of an electric go-kart include:

1. **Structural integrity**: Creating a go-kart body that withstands every form of possible pressure, including weight, a body that is torsionally rigid, impact resistant, and stress resistant at a low weight.
2. **Performance optimization**: Level of acceleration, maximum attainable speed, and go-kart handling are all factors that should be equal to or better than the performance of current ICE go-karts.
3. **Safety assurance**: Offering complete protective systems for the go-kart, such as emergency exit braking systems, crash protective systems, and systems that prevent the driver from being ejected.
4. **Energy efficiency**: Creating a go-kart that maximizes overall distance that can be driven through the use of an optimized battery to suit the motor, and includes regenerative braking.
5. **Cost-effectiveness**: Creating an optimal design of go-kart that meets the performance and safety goals while still being economically feasible.
6. **Environmental responsibility**: Consider the overall impact on the environment of all components, including the battery: emissions, and battery production, and recycling.

The development of electric vehicles (EVs) has been ongoing and has flourished in the past few years, along with the current buzz around EVs. The first pass cards issued in the late 1950s could only be used with gasoline engines. Electric motors became incorporated in the 1970s and 80s in the first wave of production for EVs. Early electric motors did not compete as well as gasoline engines due to the limitations on battery technology at the time. Improvements made during the 1990s to battery systems, such as lithium-ion and nickel hydride batteries, revolutionized the performance and productivity of electric pass karts. These devices reduced noise pollution and fuel emissions, and increased environmental sustainability and became popular during the 2000s on indoor running tracks.

In the last decade, technology improvements that allow electric cross karts to match gas models in speed and maneuverability have made these karts increasingly popular. Modern electric go-karts offer better telemetry and configurable systems and have regenerative braking systems, which improve presentation. As well as recreational purposes, these karts have now got educational applications to demonstrate sustainable electricity use and processes in the automotive industry.

Environmental and educational goals drive the creation of the EV move-kart as an alternative to traditional gas-powered karts. Incorporating a clean and efficient electric drive system provides a more positive, environmentally friendly (non-destructive) alternative to traditional gas karts. Educational outcomes surrounding advanced automotive technologies such as electric drivetrains, battery management systems (BMS), and energy optimization algorithms can be more accessible to automotive development students. Increasing demand for electric karts complements a more positive educational setting and encourages innovative and entrepreneurial pursuits to integrate further into outdoor leisure EV activities. Safety of EV systems, spectators, and drivers is the primary concern and is built into the design. EV move-kart is cost-efficient and easily deployable in various educational contexts, encouraging the production of leisure EV systems. This project not only fosters innovative educational practice in electric mobility and sustainable activities. Thanks to its multifaceted approach, it also increased educational practice in the mobility of electric systems.

12.2 ELECTRIC GO-KART FUNDAMENTALS AND DESIGN REQUIREMENTS

12.2.1 Go-kart classification and standards

The International Karting Commission, a subdivision of the Federation International Automobile, defines a go-kart as a vehicle without any bodywork, consisting of four wheels with two wheels being responsible for steering while the other pair transmits power to the ground. Engine displacement, power output, and class designation determine the respective categories that go-karting falls into. Electric go-karts, which hinge on power and voltage parameters rather than displacement categories, require different classification categories.

The four categories into which the chassis of go-karts are the following: Open chassis go-kart, which does not have a roll cage, then there is a Caged go-kart chassis which is a pre-fabricated chassis with roll cage protection. With regards to electric go-karts, cage designs are a critical element for protection against electrical risks, while also ensuring the chassis has adequate rigidity to withstand impact

12.2.2 Design specifications and performance targets

Electric go-kart specifications describe operational parameters in multiple fields, such as:

- **Performance specifications:** Maximum attained speed (40–80 km/h for recreational and educational purposes), acceleration (5–8 seconds from 0–50 km/h), handling (lateral acceleration of up to 1.5g on the plane), and ranges (should be at least 20–30 minutes of uninterrupted running at the nominal power).

- **Dimensional constraints:** Length in the region of 1500–1800 mm, width of 1000–1200 mm, wheelbase of 1000–1200 mm, ground clearance (50–100 mm), and maximum curb weight (100–150 kg excluding the driver).
- **Safety requirements:** Maximum structural impact deformation (100 mm at the bumper), rollover stability levels (1.4 g minimum static), braking performance (≤10 meters from a maximum speed), and electrical safety (50V minimum across an insulator).

12.3 CHASSIS DESIGN AND STRUCTURAL ENGINEERING

12.3.1 Material selection

AISI 1018 was deemed appropriate when compared to other materials but instead, AISI 4130 was used because of its excellent weldability, high fracture toughness, and plasticity, which helped it become a great option for the chassis [1]. The selection of materials for the go-kart chassis is a vital balancing act of the optimum structural performance, weight, price, and ease of manufacturability. The AISI 4130 chromium-molybdenum steel has a number of benefits for construction of the chassis tube as follows:

- Yield strength: About 435 MPa, which is a considerable load-carrying ability.
- Dynamic strength: For inverse bending dynamic strength (σ_d-1) is around 280 MPa and for inverse torsional dynamic strength (τ_d,t-1) is 180 MPa.
- Weldability: Very good characteristics of weldability, which allows construction of joints with great efficiency

Cost-effectiveness: Fairly accessible and reasonably priced materials
The structural demand requires the least amount of possible mass (weight) to be used which on the other hand means strength and rigidity for the assumed loads, with minimum costs for the joints (mostly welds), and standard part fittings.

12.3.2 Chassis geometry and load distribution

Chassis analysis was conducted with CAD/FEA software packages and consisted of an initial phase manipulating and optimizing chassis geometry with beam elements, and later with solid elements targeted at sub-local modeling of the optimized chassis [2]. Efficient design of the chassis incorporates thorough and systematic analyses of the different load paths, stress concentration, and the appropriate response system of the chassis at different operational states.

- Frame topology: Monocoque micro-tubular framework of the go-kart chassis, with the micro-tubular structure's primary load-bearing parts engineered/spatially arranged to gain the optimal stiffness and weight ratio. The major micro-tubular frame parts are:
- Main hoop: The primary structure for roll-over protection, extending from the back axle mount to the driver's seat area.
- Front longitudinal members: Tubular from the axle mount at the front wheels to the main hoop to resist and counteract bending loads on the structure during acceleration.
- Side rails: Arrays of longitudinal members for providing and maintaining the structure's stiffness to torsion and for absorbing impacts.
- Rear axle carrier: Aesthetic and functional subframe supporting and sustaining the motor, battery, and rear wheels.
- Steering column supports: Steer shaft and shaft column structure impact and offer protection.

Load cases: Chassis structure analysis normally examines and incorporates multiple critical loads, and the following are typical:

1. Static loading: Weight distribution including driver + chassis (approx. 60–90 kg), motor + drivetrain (15–20 kg), battery (25–35 kg), and other structural frame elements.
2. Dynamic acceleration: During acceleration and braking, there are inertial loads on the structure over the longitudinal axis, which during operation, acceleration values are typically 0.4–0.6 g.
3. Cornering loads: The side load is generated on suspension mounting points due to lateral inertial forces while turning.
4. Impact scenarios: Impact strategies on front, side, and rear collisions reviewed to European New Car Assessment Programme (ENCAP) standard with a frontal impact at 64 kmph, side impact at 48 kmph, and rear impact at 50 kmph [3].
5. Torsional loading: Bump analysis where a couple was applied on the chassis, with deformation of 1.51 mm, maximum stress of 131.41 MPa, FOS 2.35, and torsion values were found.

12.3.3 Stress analysis and safety factors

Stress distributions and deformations obtained from comprehensive finite element analysis (FEA) determine potential failure under all loading conditions. Design finite element analysis (DFEA) with design verification plan (DVP) and design failure mode and effects analysis (DFMEA) fall-back analysis was performed to enhance quality and safety parameters. Modifications were done to components with predictable loading to ensure no design failure occurred.

Stress analysis methodology:

1. Ascertaining the peak equivalent stress for each loading condition
2. Locating design areas for stress concentration and subsequently modifying geometry
3. Soderberg safety criteria for calculating dynamic strength safety factors.
4. Design-iterating to reduce peak weight and stresses.

Safety factor determination: If the front impact load is transferred to the front most then the chassis and driver is protected with F.O.S = 1.148 (material strength = yield point). Safety factors for go-kart chassis are typically 1.5 to 3.0; safety factors climb with criticality of load case and design uncertainty.

12.4 ELECTRIC PROPULSION SYSTEMS

12.4.1 Brushless DC motor technology

The virtues of brushless DC (BLDC) motors have made them popular in the automotive industry. They are more efficient, have a longer service life, have a more compact design, provide a higher starting torque, and provide a consistent torque at varying speeds, which is crucial for a vehicle's acceleration and for climbing [4]. BLDC motors, also called electronically commutated motors, are synchronous motors that convert direct current (DC) power into mechanical energy. They employ an electronic commutator, which alternates which windings of the motor are energized, to make the direction of the motor's torque vector rotate [5].

Some of the structural advantages of the BLDC motors are:

- Efficiency: The most efficient drive is achieved by replacing the ICE with the DC motor and lithium-ion battery. The power they produce is 1.6 kW and is constant in driving mode [6]. For competitive racing, a range of 1.6 kW to 5 kW motors is used. For recreational purposes, 2 to 3 kW is optimal.
- Torque density: Having very high values on torque-to-weight ratios facilitates the design of compact yet high-performing motors.
- Maintenance: The absence of brush and commutator abrasion increases the lifetime of the device and reduces costs associated with upkeep.

- Control precision: Digital control algorithms allow for precise control of the speed and torque of the motor, while electronically commutating to maintain efficiency [7].

12.4.2 Motor selection criteria

While determining the replacement motors for the electric go-karts, the following attributes were considered:

- Power rating: One of the main driving attributes of the go-kart is the replacement of the gasoline ICE and lithium-ion battery with the BLDC motor used for the driving mechanism. For the optimally calculated power train for sustained operation, the motor power should be rated at 1.6 kW [8]. For the motors adapted for competitive racing, the range is 1.6 kW–5 kW, whereas for recreational purposes, the optimal power rating is in the range of 2–3 kW.
- Torque characteristics: Motor selection must ensure adequate starting torque to accelerate the vehicle from standstill while limiting wheel slip. A BLDC motor of 2000 W capacity which delivered a torque of about 6.16 Nm, was used. Torque requirements depend on tire friction characteristics, gear reduction ratios, and acceleration targets.
- Voltage and current ratings: The research involved vehicles equipped with a BLDC motor (5 KW, 48 V, 3000 RPM) and a lithium-ion battery system (48V, 35Ah). Popular go-kart Voltage specifications range from 36V to 72V systems and in most cases 48V is the optimal compromise in terms of safety and performance [9].

Speed characteristics: Motor speed ratings (typically 2000–4000 RPM for direct-drive or 3000–6000 RPM for geared applications) must align with target vehicle maximum speed through appropriate gear reduction calculations.

12.4.3 Electronic speed control and motor management

BLDC motors, similar to other motor types, need their own specialized electronic control units for thermal management, speed regulation, and winding current commutation. The motor-blade-communicating speed controller receives three-phase electrical output from the inverter, which is fed by the EV's battery pack [10].

Control algorithms: Multiple control approaches provide varying performance and computational requirements:

- **Six-step commutation** (trapezoidal): Most basic control option with moderate performance and computational costs, where Hall effect sensors are used to develop a commutation timing strategy.
- **Field-oriented control** (FOC): More sophisticated control option where maximum efficiency at a targeted energy level is achieved by using d-q axis decomposition, albeit at elevated computational costs [5].
- **Fuzzy logic control** (FLC): FLC is a practical approach for nonlinear control systems, as it forgoes the use of complex empirical algebra for a more intuitive and computational lightweight approach. FLC is synonymous with precise control for systems where minimal speed variations need to be achieved, as it is associated with the elimination of overshoot, steady-state error, and smooth transitions [11].

12.5 BATTERY SYSTEMS AND ENERGY STORAGE

12.5.1 Lithium-ion battery technology

The wide margin of benefits, including high energy density, long lifespan, high specific and cycle efficiency, low individual discharge rates, and quick reaction time, positions lithium-ion batteries as powerful and cost-effective for driving EVs, an alternative to ICE, and facilitates global clean transport [12]. Li-ion batteries are the most effective from all rechargeable batteries for electric mobility due to low self-discharge rate, extended operational range, high energy density, and long life cycle [13].

Battery chemistry considerations: Standard lithium-ion chemistries used in go-karts include:

- **Lithium iron phosphate** ($LiFePO_4$): More than 3000 cycles of enhanced cycle life, weakening of energy density with lower thermal stability, safer chemistry [14].
- **Nickel–cobalt–aluminum** (NCA): Higher energy density (200–260 Wh/kg) enabling extended range, moderate cycle life (1000+ cycles) [15].
- **Nickel–manganese–cobalt** (NMC): Balanced characteristics with good energy density (150–220 Wh/kg) and cycle life (1000–2000 cycles) [16].

12.5.2 Battery pack configuration and sizing

When determining the battery capacity, there are a few factors that need to be taken into account.

Energy requirements: Depending on the weight of the vehicle, the efficiency of the driving system, the terrain, and the local environmental conditions, the total system energy consumption will change. Considering a go-kart of 120 kg (60 kg of that being the driver) and an 85% efficient drivetrain, the energy requirements can be calculated as follows:

Energy requirement = (Vehicle weight + Driver weight) × Average acceleration × Distance / Drivetrain efficiency

Analysing the Li-ion battery capacity needs will help optimize the performance of the vehicle and help define a target battery capacity, which must offer a driving time of 20 minutes.

- **Pack architecture:** A battery pack with a BMS that manages the state of the pack, controls/powers the pack, and manages the interconnections of the pack to the EV components. Go-kart battery configurations use series and parallel connections as follows:
- **Series configuration:** Cells are placed one after the other in a row to achieve the required voltage in a pack. (e.g. to achieve 48V, one would need 13 or 14 cells of 3.7V nominal lithium-ion cells)
- **Parallel configuration:** A few series configurations are placed in parallel to achieve a higher total current capacity and a lower current stress on the cells.
- **Module assembly:** Cells are put in a case that has a cell balancing circuit, thermal management control, and other structural support.

12.5.3 Battery management systems

A BMS caters to the needs of each battery one at a time so that each one can function at the same level, thus synchronizing operations, prolonging the battery life, and safely securing its use [17]. BMS must have vital functions, such as voltage, temperature, and current monitoring, state of charge (SoC) of a battery, and balancing of Li-ion batteries [18].

1. **Voltage monitoring:** Continuous measurement of each of the cell's voltages as well as the voltages of the modules and assembled packs to identify potential problems and prevent over/under voltage conditions
2. **Temperature management:** Monitoring the temperature of the battery pack to prevent excessive temperature increases, then charge and discharge rates are modified to match the temperature [18].
3. **Current monitoring:** Measurement of the discharge and charge current of the pack to determine SoC and to identify overcurrent [14].

4. **Cell balancing:** Active or passive equalization of individual cell voltages to maintain uniform voltage distribution.
5. **State estimation:** Through coulomb counting, SoC, State of Health (SoH), and remaining life of the battery are calculated, through monitoring and mapping the open circuit voltage, and using a Kalman filter [18].
6. **Protection:** Overvoltage, undervoltage, overcurrent, overtemperature, shorts, and other fault conditions will automatically disconnect the circuit controlling the charge and discharge.

12.6 TRANSMISSION AND DRIVETRAIN SYSTEMS

12.6.1 Transmission architecture

Electric go-karts utilize a range of different transmission systems, each with unique benefits and disadvantages:

- **Direct drive:** Highest efficiency with no interceding gearing, but may lack the needed RPM range and require a high RPM motor to achieve high speeds [19].
- **Single-speed reduction:** Gearboxes with fixed-unit ratio offer the best degree of torque multiplication for acceleration. The drive shaft on the rear wheels is designed for higher torque, in case a more powerful motor is installed [20].
- **Chain drive:** These systems are light, low maintenance, and highly efficient. The ratio of transmission can be flexibly modified with a sprocket change. However, the systems do lose some efficiency to friction [21].

12.6.2 Gear ratio calculation

The maximum achievable speed and the acceleration of the go-kart must be taken into account for the best-suited gear choice:

- **Torque at wheels** = Motor Torque × Gear Ratio × Efficiency; **Vehicle velocity** = (Motor RPM × 60 × Tire Circumference) / (Gear Ratio × 1000)
- For a typical electric go-kart with 6 Nm motor torque, target 50 km/h maximum speed, 300 mm tire radius, and 2000 RPM motor speed:
- Required gear ratio = (Motor RPM × Tire Circumference) / (Vehicle Velocity × Gear Ratio). Approximately 4:1 to 5:1 reduction provides a suitable balance between acceleration and top speed.

The rotating wheel normal forces result in normal stress components that change harmonically, while constant shear stress components caused by torque result in a disproportional state of stress that Soderberg's criteria of dynamic strength takes into design.

12.6.3 Drive shaft design and analysis

The drive shaft was documented in a calculation based on the principles of dynamic strength analysis, which is something that is not present in a lot of the listed references. The design of the drive shaft must include the design of the shaft to include bending and torsional stresses at the same time because of dynamic loading.

- **Torsional stress analysis:** Maximum shear stress from torque transmission: $\tau = T \times r / J$, where T = applied torque, r = shaft radius, J = polar moment of inertia
- **Bending stress analysis:** There is a combination of stresses due to lateral loads while cornering and during acceleration: $\sigma\_b = M \times c / I$, where M = bending moment, c = distance from the neutral axis, I = the second moment of inertia.

The harmonic changes of normal stress components due to the wheel normal forces and the constant shear stress components due to torque result in a disproportional state of stress that Soderberg's criteria of dynamic strength considers in design.

12.7 STEERING AND SUSPENSION SYSTEMS

12.7.1 Steering geometry and control

The go-kart steering systems do not include suspension as compared to cars. The steering system included Ackerman angle, while the caster angle and kingpin inclinations were roughly 12° and 7°, respectively

- **Ackermann steering geometry:** Guarantees that the four wheels encircle concentric circles at one vertex when making a turn. This development lowers tire slip and better handle predictability. The design has steering tie rods that are set at angle, producing different steering angles for the inside and outside wheels, based on the cornering radius.
- **Caster angle:** Is anywhere from 7° to 12° of angle at the steering axis to provide self-centering forces, maintaining stability for the vehicle on a straight path, and restoring control after the vehicle has been disturbed.

12.7.2 Chassis compliance as suspension

- Go-kart frames consist of tubular steel, which is stiff and flexible. Their design allows them to act as suspension components while being durable enough to withstand the impacts resulting from the chassis maneuvers. Additional features of the chassis, such as tube diameter and wall thickness, determine key features of the chassis, including vertical compliances and twisting configurations.
- **Vertical compliance:** Thinner walls and smaller-diameter tubes allow for improved vertical flexing.
- **Torsional rigidity:** Larger tubes and specific configurations of the chassis geometry increase the resistance of the chassis to twisting.
- **Damping:** Absence of friction allows the material and the joints to provide passive damping of any oscillations. A chassis with a high rigidity level will perform best in dry conditions, whereas a flexible one in wet conditions.

12.8 BRAKING SYSTEMS

12.8.1 Hydraulic disc brakes

In order to reduce the rotational speed of vehicle wheels, vehicle wheels are equipped with hydraulic disc brakes that create friction to slow the vehicle down, removing energy in the form of heat, which flows through the rotor vents to the environment.

Braking system architecture: Hydraulic systems typically use:

- Master cylinder: 10 mm of master cylinder diameter, 25.4 mm of caliper piston diameter, 4:1 brake pedal lever ratio, 2.237 m of stopping distance.
- Brake lines: Dual circuits that increase redundancy and safety margin.
- Calipers: Brake pads are pressed against the rotor surfaces by means of hydraulics.
- Rotors: Discs that are either ventilated or are solid, which are used to absorb and dissipate energy through the brakes

12.8.2 Regenerative braking

The ability to recover energy while slowing down due to electro propulsion system (EPS) is achieved by regenerative braking and deceleration. The system is equipped with an electronic differential, which enhances the efficiency of the EV driving system, and the energy feedback to the battery is improved through the BLDC motor, especially when the vehicle is in braking mode. This improves the energy economy and prolongs battery life. Regenerative braking captures roughly 10–25% of kinetic energy based *on the deceleration rate, motor* efficiencies, and battery conditions. Its incorporation includes the following:

- Bidirectional motor controller: Necessary for controlling the motor as a generator during regen cycles.
- Electrical isolation: Preventing unwanted discharges through the motor's windings when the motor is a generator.
- Blended braking algorithm: Integrates regen and friction braking in a manner that preserves pedal feel and effort.

12.8.3 Emergency braking and safety systems

Safety design includes the provision of a kill switch and front, rear, and side bumpers to reduce collisions and protect the driver.

12.9 DESIGNED METHODOLOGY

12.9.1 Frame design materials

The main objective of the chassis was to efficiently and safely connect every component needed for the vehicle, including the driver. The plan was guaranteed:

(i) Driver safety.
(ii) Gravity is still small.
(iii) Light and strong chassis
(iv) Each part was securely installed on the chassis.

AISI 1020 is composed of the following substances.

- Carbon C equals 0.18%. Iron Fe = 99.20 per cent, Manganese Mn = 0.4%. Porphyrius P = 0.03%. Sulfur content S = 0.04%.
- The EV go-kart materials specifications are mentioned in Table 12.1, which provides certain parameters to stay with.

When the roll-cage element was originally introduced, safety regulations were ignored. Maintaining an optimal distance between driver body breathing and complex elements such as the engine compartment and firewall was a key safety criterion prioritized during design, and other safety measures were implemented to meet the prerequisites it has been dealt with. The chassis design was aimed at providing occupants with more space for

Table 12.1 Electrical vehicle go-kart materials specification

S. No	*Parameters*	*Specification*
1	Solids	7.87 g/cc
2	Solids, Brinell	121
3	Hardware, Noop	140
4	Solids, Rockwell	B68
5	Solids, Vickers	120
6	Tensile strength, ultimately	420 MPa
7	Tensile strength, yield	350 MPa
8	Expansion at rest	15%
9	Local reduction of	40%
10	Modulus of elasticity	186 GPa
11	The maximum rate is	148 GPa
12	Poisson ratio	0.29
13	Machinery	65%
14	The viscosity is	72.0 GPa

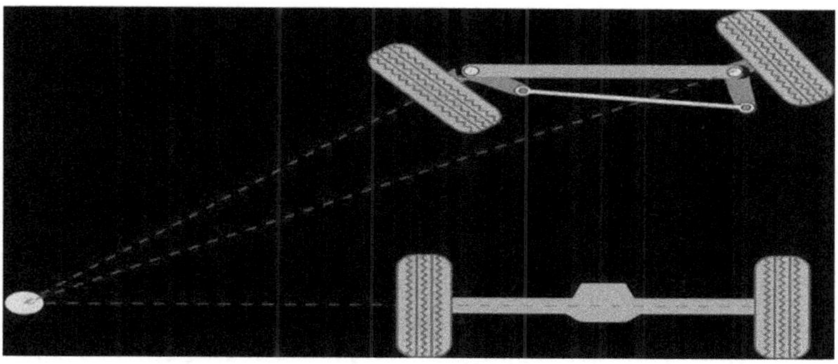

Figure 12.1 Concept of Ackerman steering mechanism.

faster maneuvering. The fire extinguisher is conveniently positioned, with thick foam mounted on the driver's side tanks.

12.9.2 Steering system configuration

Due to its simple design, instrumentation, and 1:1 steering ratio, this electric go-kart, which uses the same mechanical system options in this steering wheel, uses an Ackerman steering wheel as its steering system.

Ackerman's 60° lock-to-lock rudder wheel turns are ideal for tracking because they allow fast twists with fine detail while maintaining great accuracy (Figure 12.1).

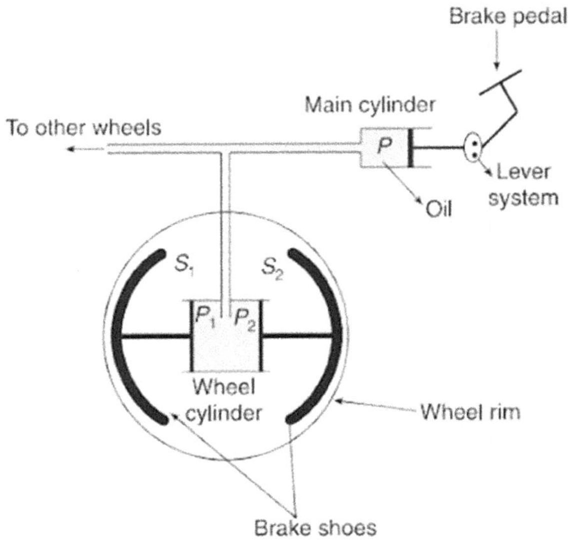

Figure 12.2 Breaking system.

12.9.3 Brake design parameters

Despite its thickness and greater durability, it is a small pass-kart scale (Figure 12.2).

Brake calipers inject hydraulic fluid, where pistons apply pressure to the brake pads on the brake discs or rotors attached to the rear axle. Contraction creates friction, which in turn slows the rotation of the cross kart's rear tires. The brakes create friction as they press on the discs, slowing the go-kart to an effective stop. This process also converts the go-kart's kinetic energy into heat energy (Figures 12.3 and 12.4).

Also, little maintenance is required, and good control modulation can be provided for controls of induction strength.

12.9.4 Electrical system

Especially in terms of electrical safety and shock risk, the 48V system is practical and safe for the driver, providing adequate power

Furthermore, the motor's brushless architecture offers many advantages over conventional brush motors, such as increased longevity, less maintenance, and higher efficiency. These characteristics are important to reduce when it is released to ensure continuous operation of the race.

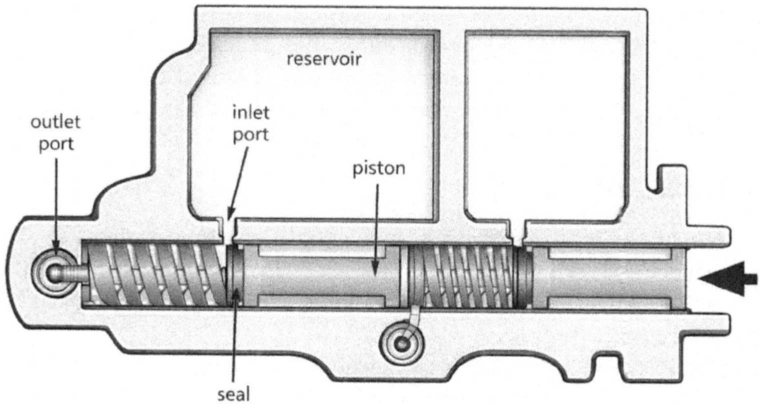

Figure 12.3 Master cylinder of hydraulic braking system.

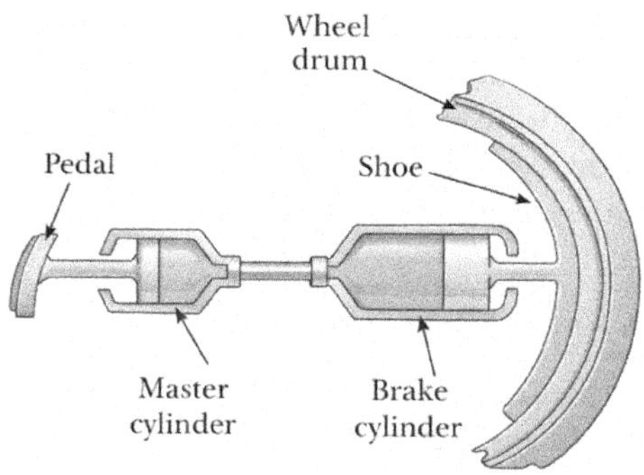

Figure 12.4 Brake pedal.

12.10 DESIGNING PARAMETERS FOR GO-KART

Electric propulsion engine: The aim of the research is to provide rapid, high torque, and economical power consumption, which is more dependent on the choice of electric propulsion. The electric motor plays an important role in defining standard performance, all of which electric pass-karts are and into the heart.

12.10.1 High-performance EVs motor selection

Choosing the right electric motor is important for an electric pass kart system. The selected motor should maintain high torque levels while providing the power required for fast acceleration. Doing this requires careful consideration of the mechanical characteristics of the work, such as strength, torque, and force orientation. Those requirements are met with state-of-the-art electric motors. The power output of the cross-kart was chosen to reach aggressive speeds, giving drivers an enjoyable experience. A higher torque motor is also needed to accelerate faster and more efficiently, making it easier for the go-kart to handle rough terrain and tight corners.

12.10.2 Battery age

The layout of the electrical pass kart is equally as important as the choice of battery technology. To maximize the electrical-weight ratio of the kart, which directly affects its speed and weight, the electrical potential of the weight must also be balanced.

Because of their high electrical density and light weight, lithium-ion batteries are generally preferred. These batteries are ideal for powering pass karts because they combine a good amount of garage power with power output. Also, lithium-ion batteries are known for their longevity and reliability, assuring that go-karts will last longer without frequent battery changes, to maximize go-karts diversity and performance, the capacity of the battery % is selected to provide enough power at the same time as maintaining a reasonable load with possible battery management system (BMS) is integrated for monitoring and monitoring battery health, temperature and charge cycles, ensuring safe and efficient operation.

12.10.3 Chassis and safety functions

The chassis is the structural basis of the electrical pass-kart, and its design has a major impact on the functionality, stability, and safety standards, combined with its implications in chassis design.

12.10.4 Top chassis layout

This layout focuses on achieving better weight distribution and stress levels. At some point during a fast spin, stabilization is required due to the uniform weight distribution of the go-kart chassis. While reducing specific weight, the chassis must provide adequate support for batteries, motors, and other components. Careful consideration must be given to material choices to achieve the best balance of weight and structural integrity, which may include lightweight but durable alloys or composites. Computer-aided design (CAD) and FEA are used to design and optimize chassis structures for stiffness and strength.

12.11 CHASSIS DESIGN

The vehicle chassis structural support determines all its functions and fulfills its characteristics. It is the cornerstone of every car. In the illustration of our electric pass kart roadmap, chassis design has proven to be key to achieving our goals of efficiency, balance, and safety.

- **Design concerns:** Before we get into the details of chassis design, it is important to briefly discuss the key factors that influenced our decision-making process:
- **Weight distribution:** Weight is distributed as evenly as possible to ensure mobility and stability.
- **Fabric selection:** Choosing the right materials that balance strength, weight, and cost.
- **Improving models:** Building models to assess the feasibility of policy ideas and identifying areas for increased capacity.
- **Testing and development:** Comprehensive testing of the overall performance of the chassis in real- world conditions and improvement of the design based primarily on feedback and statistical analysis.

Ultimately, chassis design is central to our electric cross-kart efforts, encompassing our commitment to safety, performance, and innovation. Through innovation, we have not only met rigorous technical standards but the expectations we had in building this platform exceeded (Figures 12.5, 12.6, and 12.7).

12.12 CONCLUSION

The design and implementation of an electrifying move kart that prioritizes performance and sustainability is a major breakthrough in recreational vehicle sport. This research effort provides evidence that electric propulsion is capable of reporting if it is fun for on the ride while fulfilling environmental

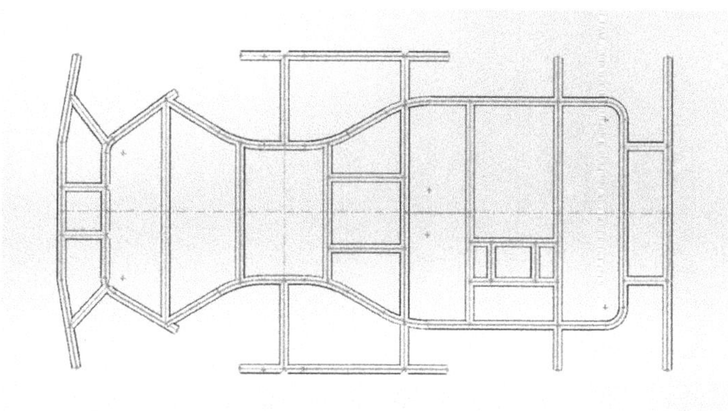

Figure 12.5 (a) and (b) 2D model of chassis design with dimensions.

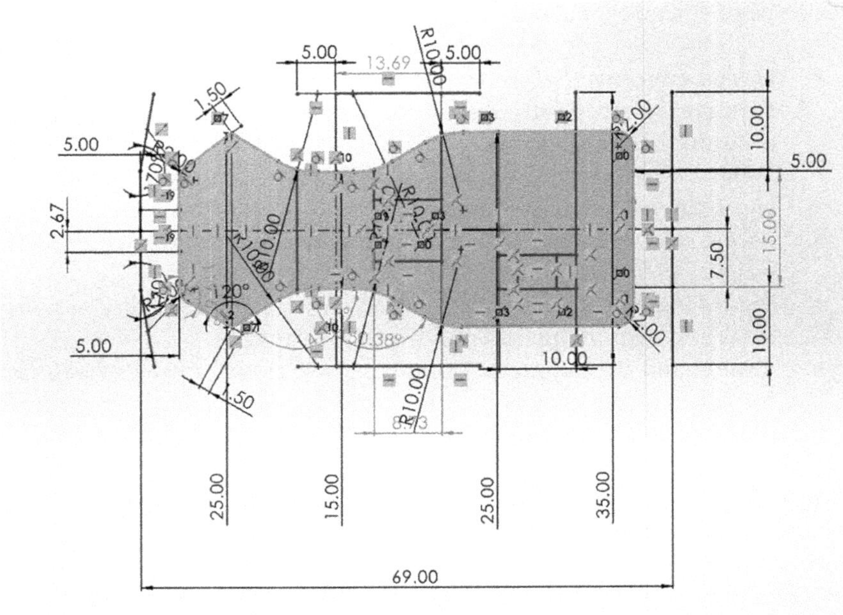

Figure 12.5 (Continued)

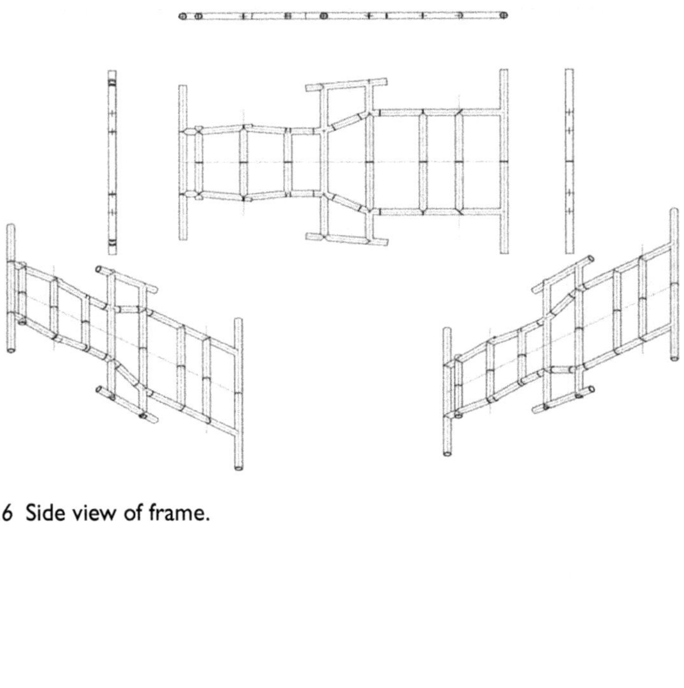

Figure 12.6 Side view of frame.

Figure 12.7 Assembled model.

sustainability and electrical performance standards. The successful integration of state-of-the-art technologies has significantly improved system performance under electronics.

The electric pass-kart project demonstrates the potential of electric propulsion in recreational motorsports. This research shows that an electric pass kart can not only match the performance and appeal offered by their fossil gas-powered competitors, which have historically been fuel-powered machines, with fast speed, great power, and exceptional handling. Electric go-karts were a challenging fantasy, paving the way for a whole new era of exciting motor racing journalism.

Data Availability Statement: In this article, data sharing is not applicable, because in this study, no new data is created or analyzed.

REFERENCES

[1] Chohuan, N., Porwal, A., & Chouhan, K. (2020). Design and fabrication of electric go-kart. *International Journal of Engineering Research and Technology (IJERT)*, 9(9), 659–671.

[2] Tao, L., Zhang, S., & Li, S. (2020). Finite element analysis and optimization of electric go-kart chassis structure. *Journal of the Institution of*

Engineers (India): Series C, 101(3), 517–527. https://doi.org/10.1007/s40030-020-00455-3

[3] Gillespie, T. D. (1992). Fundamentals of vehicle dynamics. Society of Automotive Engineers, Warrendale, PA.

[4] Chau, K. T. (2015). Electric vehicle machines and drives: Design, analysis and applications. Wiley, Singapore. ISBN 978-1-118-75252-4.

[5] Schaltz, E. (2011). Topologies and driving concepts for permanent magnet synchronous motor driven electric vehicles. *IEEE Transactions on Industrial Electronics*, 59(10), 3799–3809.

[6] Ehsani, M., Gao, Y., & Gay, S. E. (2005). Modern electric, hybrid electric, and fuel cell vehicles: Fundamentals, theory, and design. CRC Press, Boca Raton, FL.

[7] Miller, J. M. (2006). Propulsion systems for hybrid vehicles (2nd ed.). Institution of Electrical Engineers, London, UK.

[8] Arshad, M. F., Ahmad, M. F., Khalid, A., Ishak, I. A., Abidin, S. F. Z., & Razali, M. (2022). Investigation the influences of electric motor and electric go kart performance. *Journal of Automotive Powertrain and Transportation Technology,* 2(2), 17–23.

[9] Vinod, K. P., & Kannan, N. (2018). Design and fabrication of an electric go-kart with battery management system. *Advances in Mechanical Engineering*, 10, 1687814018766651.

[10] Mohan, N., Undeland, T. M., & Robbins, W. P. (2003). Power electronics: Converters, applications, and design (3rd ed.). Wiley, New York.

[11] Lukic, S. M., Cao, J., Bansal, R. C., Rodriguez, F., & Emadi, A. (2011). Energy storage systems for automotive applications. *IEEE Transactions on Industrial Electronics*, 55(6), 2258–2267.

[12] Pollet, B. G., Staffell, I., & Shang, J. L. (2012). Current status of hybrid, battery and fuel cell electric vehicles: State of the art paper. *Journal of Power Sources,* 239, 687–713. https://doi.org/10.1016/j.jpowsour.2012.11.144

[13] Bhatt, B., Sharma, H., Arora, K., Joshi, G. P., & Shrestha, B. (2023). Levy flight-based improved grey wolf optimization: A solution for various engineering problems. *Mathematics, 11*(7), 1745. https://doi.org/10.3390/math11071745.

[14] He, H., Xiong, R., & Fan, J. (2011). Evaluation of lithium-ion battery equivalent circuit models for state of charge estimation by an extended Kalman filter. *Energies*, 4(4), 582–598. https://doi.org/10.3390/en4040582.

[15] Wang, J., Ye, Y., Pan, R., Gao, T., Zhang, C., & Husain, I. (2018). High temperature calorimetric measurements of lithium ion battery charging and discharging under wide current range. *Journal of Power Sources,* 374, 22–31. https://doi.org/10.1016/j.jpowsour.2017.10.077

[16] Zhang, S. S., Xu, K., & Jow, T. R. (2006). The low temperature performance of Li-ion batteries. *Journal of Power Sources*, 115(1), 137–144. https://doi.org/10.1016/S0378-7753(02)00618-3

[17] Singh, P., Arora, K., & Rathore, U. C. (2022). Control strategies for improvement of power quality in grid connected variable speed WECS with DFIG–An overview. *Journal of Physics: Conference Series, 2327*(1), 012008. https://doi.org/10.1088/1742-6596/2327/1/012008

[18] Junghans, G., & Sonnenschein, R. (2018). Battery management systems for electric vehicles: Design principles and implementation strategies. *IEEE Transactions on Industrial Electronics*, 65(12), 9560–9570.
[19] Rong, X., Tao, H., & Guo, S. (2014). State-of-charge and state-of-health co-estimation of lithium-ion batteries based on a novel combined model. *IEEE Transactions on Industrial Electronics,* 62(12), 7874–7882.
[20] Prabhakar, H., & Karthik, S. (2019). Design, analysis, and testing of a formula student electric vehicle drivetrain. *IEEE Access*, 7, 34420–34431.
[21] Masrur, M. A., Soylu, S., & Sreeramareddy, D. (2007). Powertrain design for heavy hybrid electric vehicles. *IEEE Transactions on Vehicular Technology*, 56(3), 1137–1149.

Index

For Product Safety Concerns and Information please contact our EU
representative GPSR@taylorandfrancis.com
Taylor & Francis Verlag GmbH, Kaufingerstraße 24, 80331 München, Germany

www.ingramcontent.com/pod-product-compliance
Ingram Content Group UK Ltd.
Pitfield, Milton Keynes, MK11 3LW, UK
UKHW021505240926
475702UK00035B/436
* 9 7 8 1 0 4 1 1 7 3 7 8 6 *